THE BOOK OF
THE MOON

www.**rbooks**.co.uk

THE BOOK OF THE MOON

Rick Stroud

Doubleday

LONDON · TORONTO · SYDNEY · AUCKLAND · JOHANNESBURG

TRANSWORLD PUBLISHERS
61–63 Uxbridge Road, London W5 5SA
A Random House Group Company
www.rbooks.co.uk

First published in Great Britain
in 2009 by Doubleday
an imprint of Transworld Publishers

Copyright © Rick Stroud 2009

A CIP catalogue record for this book
is available from the British Library.

ISBN 9780385613866

Addresses for Random House Group Ltd companies outside the UK
can be found at: www.randomhouse.co.uk
The Random House Group Ltd Reg. No. 954009

The Random House Group Limited supports The Forest Stewardship
Council (FSC), the leading international forest-certification organization. All our
titles that are printed on Greenpeace-approved FSC-certified paper carry the FSC logo.
Our paper procurement policy can be found at
www.rbooks.co.uk/environment

Typeset in Bulmer and Ocean Sans
Designed by Bobby&Co, London
Printed and bound in Germany by GGP Media GmbH

2 4 6 8 10 9 7 5 3 1

Contents

Acknowledgements

The moon has many fans and I have met a lot of them in the writing of this book. They were all encouraging and keen to help, feeding me with articles and information. I would like to thank every one of them, especially and in no order: Harry and Charlie Cory Wright, Ariane Koek, Arzu Tahsin, Sarah MacLachlan, Kit Hesketh Harvey, Matthew Rice, Tanaz Eshagian, the staff of the British Library, especially Science 3, the staff of the London Library, Mark Pringle, David Morrissey, Rupert Lycett Green, Jonathan Powell and Kerry Lee Crabbe.

I would also like to thank friends who were very generous with their time and expertise. In particular: David Nash who read the 'Facts and Figures' chapter and made many valuable suggestions; A. C. Grayling for his advice and all the articles he sent me; Rosemary Hill who explained Stonehenge; Lynette West at the Biodynamic Education Centre for her advice about lunar gardening; Elizabeth Sheinkman and her lunar sister; everybody at Ruskin Hill Farm and especially Bernie Courts; Mario Forte, Joanna Fabijanczyk and Lina Azlauskaite at the Star Café for daily encouragement; Lauren Kassell for her tutorial on the Occult; Jenny Uglow for her advice on the Lunar Men; Susannah Clapp who never failed to pass on the lunar information she encountered in her travels; Michael Eleftheriades for sharing his internet expertise; Laura Pumphrey for an early steer on geology; Geraint Evans who lent me films and read the chapter on astronauts; Jeremy Hardie for his thoughts on the tides; Kitty Aldridge who also lent me invaluable moon material; Ben Schott who spent hours talking to me about the project and came up with many useful ideas; and Sue Charkin who translated the Emerald Tablet.

I would like to thank my daughters Nell Gifford and Clover Stroud for reading much of the manuscript and giving helpful notes; also my grandson Jimmy Joe Hughes for his encouragement and the loan of research materials from his own collection; and my brother and co-pilot, Tim Stroud. I owe a special debt to Justine Picardie who first suggested I write this book. I would like to thank Joanna Briscoe for her encouragement and advice and many fun lunches in the British Library. Melissa Chinchillo at Fletcher & Parry in America. George Gibson at Walker Books in New York for his wise counsel. Everybody at Transworld, especially Vivien Garrett, Aislinn Casey and Sheila Lee who researched the pictures; Mark Handsley for his painstaking and rigorous copy-edit; Elizabeth Dobson and Judy Collins for reading the proofs; the book's designer Bobby Birchall and my brilliant and patient editor Katie Espiner. Finally I owe a huge thank you to everyone at A. M. Heath and especially to my agent Victoria Hobbs who read my first thoughts about the moon and championed them in the world. Any mistakes in this book are entirely my own.

Rick Stroud
HB Veronica
London

For my wife and *sine qua non* – the one and
only Alexandra Pringle

Introduction

I first became interested in the moon as a small boy growing up in the suburbs of South London. My brother and I converted our dad's small garden shed into a moon rocket. We nailed bits of wood to the walls – these were the levers with which we would control our flight. Above the levers we chalked the instruments we would use to monitor every aspect of the rocket's performance and which would be vital to our survival in the event of an emergency. We made space helmets out of cardboard boxes and every weekend we climbed into the shed, sealed the airlock door, strapped ourselves into the pilots' seats and began the tense countdown procedure that precedes any moon mission.

As an adult I read Norman Mailer's epic account of Apollo 11 and the first moon landing, *Of a Fire on the Moon*. As he described the Eagle settling on to the Sea of Tranquillity I shook with excitement. When Armstrong took that small step for a man I was right there with him, venting the oxygen from my cardboard helmet. I recently returned to Mailer's book and got the same rush.

For much of my life I have been a film director and in that role the garden shed in West Norwood became a much bigger shed on the Isle of Man. In it we built *Unity*, a lunar orbital space station, and spent six months filming the adventures of its crew for a satellite TV series called *Space Island One*.

Not all of my work has been in fiction. The rocket anorak in me was fed when I got the chance to research and make documentaries for British industry. I tackled fibre optic cable for the Ministry of Defence, domestic approach radar for British Aerospace and, perhaps best of all, a film on Sea Wolf Rocket Control Radar for Marconi Defence Systems. With Marconi I discovered a concept that still gives me a frisson of pleasure – millimetric radar.

I once even touched the moon. I worked with a film producer who had a necklace into which was set a tiny piece of moon rock, given to her when she worked on a film about NASA.

Now I live on a boat and the very water I float on is subject to the massive gravitational influence of the moon.

The moon excites me and it was to capture that excitement that I set out to write *The Book of the Moon*. I wanted a volume packed with technical detail, lists, diagrams and drawings. A book full of the derring-do of the astronauts as they juggle their rockets across the quarter of a million miles that separates us from our nearest extra-planetary neighbour. A book that I might have had as a manual as we waited for lift-off in the shed.

But I also wanted a book for my other self. A book that tells me about the mysterious, strange side of the moon. The moon that hovers above my boat and whose power is so colossal and subtle and playful it makes the Thames disappear twice a day. A disappearance that has bred in me a strong affinity with our ancestors. I know what they must have felt like 30,000 years ago as they huddled by the same disappearing river. I know that they must have stared at the vanishing water and thought, 'Is that it? Where has it all gone? Is it ever coming back?', and then staring up at the same moon I stare at they must have wondered, 'Who's in charge up there? Who do we have to placate to bring back the water?'

As I wrote I discovered that the moon does make us mad, though perhaps not in the way that we assume. It is difficult to prove that the full moon has any effect on us at all and the fact that the moon has power over the tides does not mean it has power over the water in our bodies. But mankind does extraordinary things to chart the moon's progress. Our ancestors spent millions of hours and thousands of years watching and recording what the moon does. They hauled thousands of tons of rock vast distances to construct megaliths that are in part lunar observatories. And we have done the same thing. There was no need to send a man to the moon but we just had to do it. Landing on the moon came to obsess the two most powerful nations on earth and we spent colossal fortunes and more millions of man-hours getting there. And it was, undeniably, exciting.

Today the moon presents us with a moral challenge. We are planning to go there again. We want to use the moon as a stepping stone to the stars and this time we are going to exploit its mineral resources. Both these things will completely change what the moon is. In exploiting it we will trash its unique environment. The moon is the next rainforest. Are we gripped in a new madness that will leach colossal resources into thin air at a time when we cannot get clean water to every citizen of the planet earth? This problem grew in importance for me as I wrote the book. But a bit of me still loves the excitement and the challenge of getting back on the moon. I share the madness.

So perhaps I really wrote *The Book of the Moon* for the nine-year-old me back there in the shed: the Command Module pilot anxiously scanning the dials and issuing last-minute orders to his co-pilot as the final seconds of countdown tick away. Tensing for lift-off he awaits the titanic forces that will hurl him through the suburban sky into the black depths of space and the beginning of his epic journey to that strange, frightening and exhilarating planet: the moon. I hope it will see him safe through.

Chapter 1
Facts and Figures

Some Basic Moon Facts

Mean distance from earth	**240,000 miles**
Distance from earth in light years	**1.5 light seconds**
Length of lunar day	**27.3 earth days**
Radius	**1080 miles**
Circumference	**6790 miles**
Weight	**81 quintillion tons**
Surface temperature (day)	**134°C**
Surface temperature (night)	**−154°C**
Gravity on surface	**0.1667g (1/6th of earth's)**
Orbital speed	**22,887 mph**
Driving time by car at 70 mph	**135 days**
Flying time by rocket	**60 –70 hours**
Walking time at 3 mph	**8.6 years**
Number of men who have walked on the moon	**12**
Age of oldest collected rock	**over 4 billion years**
Weight of rocks collected by Apollo missions	**842 pounds**
Widest craters	**140 miles in diameter**
Deepest craters	**15,000+ feet**
Highest mountains	**16,000 + feet**
Albedo (the extent to which an object can reflect light; see p.19)	**0.12**

The moon was formed about four and a half billion years ago. It is a satellite of the earth and is about one quarter of its size. Over the last 5 billion years, the moon and the earth have developed a complex physical relationship. That relationship is controlled by the most fundamental and powerful force in the universe: gravity. The tune gravity calls is constantly changing. The dance started with the catastrophic collision between the earth and a Mars-sized body that created the moon. It has been going on ever since, and will continue until the sun dies.

GRAVITY

The strength of the moon's gravitational field is about one sixth that of the earth's, and the field itself is uneven. Anomalies from the radio signals transmitted by the satellite Lunar Orbiter led to the discovery of patches of unexpectedly high gravity on the moon's surface. These are known as 'mascons', or 'mass concentrations'. They are mainly clustered in the northern hemisphere on the moon's near side. The largest disturbances occur in the moon's 'seas', the *Maria*, where there is lava from volcanic activity. High concentrations of mascons are to be found in the Mare Serenitatis and the Mare Imbrium. Nobody knows for certain what causes them: they may be caused by fragments of asteroid buried deep in the lava or they may be caused by the lava itself. Paradoxically, while some mascons occur where there are thick layers of lava from volcanic activity, not all areas of volcanic activity have mascons. In the early days of the Apollo missions, mascons would confuse spacecraft instruments. Bewildered astronauts wondered why their equipment displayed readings of 200 yards when the real distance was 20 yards. Mascons also destabilize the orbits of man-made satellites as they circle the moon.

In 2011, NASA plans to launch the Gravity Recovery and Interior Mission Laboratory (GRAIL). The mission will use two satellites to orbit the moon, measuring its gravitational fields in minute detail and gathering more information about the mysterious mascons.

ATMOSPHERE

The moon exists in a near perfect vacuum but it does have an atmosphere. This is created by a combination of phenomena, including solar wind, radioactive decay in the lunar surface, and possibly the influence of ultra-violet light. The moon's atmosphere changes in density from day to night, and its average weight is about 10 tonnes. Its major components are neon, hydrogen, helium, argon and sodium. The moon's atmosphere escapes partly because the moon's gravity is too weak to retain it and partly because some elements related to solar wind strip it away. The atmosphere is renewed about once every four months. There is a theory that at some time in the past 100 million years the moon had a much denser atmosphere. Astronauts found less evidence of micrometeoroid impact in the lunar soil than would be expected, which suggests that the moon had an atmosphere dense enough to burn up meteoroids before they reached the surface.

The lunar atmosphere is very fragile. Lunar exploration has already made a huge impact on it. The instruments deployed by the Apollo missions to measure atmospheric gases were swamped by the exhaust pollution from the Lunar Excursion Module and the astronauts' space suits. Each lunar landing has left pollution as great in mass as the entire existing lunar atmosphere. Some of the polluting gases took months and even years to disperse. The future could present a very serious problem. If the mass of the moon's atmosphere were to increase one hundred fold it would become stable and long-lasting. The planned mining of the lunar surface using small one-kiloton atomic devices (the equivalent of about 1000 tons of TNT) would quickly produce just such a change – and would destroy the moon's unique near-vacuum environment.

RELATIONSHIP OF THE MOON AND THE EARTH

Gravity dominates the moon's relationship with the earth. The moon's gravity pulls on the earth, and the earth's gravity pulls on the moon. Although the moon appears to orbit the earth, the two bodies

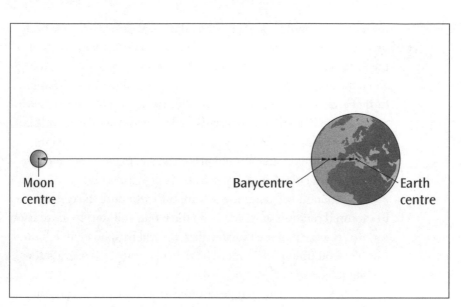

The barycentre: the common point round which the earth and the moon both rotate.

are actually orbiting a common point known as the barycentre which is situated about 1000 miles below the earth's surface.

The gravitational pull of the earth is much stronger than the gravitational pull of the moon. This causes two things to happen:

▶ It slows the rotation of both the earth and the moon.
▶ It makes the moon move away from the earth (see 'The Moon and the Earth's Tides', p. 18, for an explanation of this).

The moon has slowed to the point where it no longer rotates relative to the earth, which is why it always shows the same face. This is called captured or synchronous rotation. The same thing is happening in reverse. The moon's gravity is slowing down the rotation of the earth. In time, the earth will stop revolving relative to the moon and will show the same side to it too. The earth's day will be a month long, and that month will last for the equivalent of forty of our current earth days. The rate at which the moon slows the earth has been estimated at 0.0002 seconds per day per century.

Four billion years ago, the moon and the earth were much closer together. The Soviet unmanned and the Apollo manned missions left mirrors on the moon's surface. Every day since then lasers have been fired at the mirrors and the time taken by the reflected light to reach the earth has been recorded. Today the interval is longer than it was when the mirrors were put in place. This means the mirrors have got further away. Since Neil Armstrong and Buzz Aldrin first stepped on to the moon, it has moved 1.6 yards further away from the earth. The gap grows by about 1.5 inches per year. This is faster than the rate 2 billion years ago, when the moon is estimated to have been moving away at a speed of 0.5 inches per year.

The moon will continue to leave the earth until the earth too is in captured rotation, at which point the moon will start to move back towards the earth. It is estimated that this will happen in 50 billion (50,000,000,000) years' time. The return journey will take another 50 billion years.

These huge slow changes are known as tidal evolution. They are a physical manifestation of gravity's embrace. We see that embrace every day in the form of tides.

THE MOON AND THE EARTH'S TIDES

The side of the earth facing the moon bulges towards it. At the centre of the earth, the power of the moon's gravity is zero. On the side of the earth furthest from the moon, the power of the moon's gravity is less than zero, and the earth bulges away from the moon. This power affects everything, but can be seen most clearly in what it does to the seas. Water facing the moon is pulled towards it, while on the opposite side of the earth, water is pulled away from the moon. This results in two bulges of water on opposite sides of the earth, which we experience as high tides. The bulges follow the moon, but because the earth is rotating faster than the moon, the bulges travel just ahead of the moon rather than directly under it. The water itself has gravity, and that gravity in turn has power over the moon. The gravitational force of the bulge causes a fractional acceleration in the moon's orbit. The moon is not tethered to the earth, so as its orbit gets faster, the radius of its orbit gets bigger. This phenomenon was discovered by Johannes Kepler in the seventeenth century.

The sun too has power over the tides. When the sun, the moon and the earth are in line, a condition known as syzgy, the gravitational power of the three bodies is strong, and high tides are at their highest. These are known as spring tides and they happen when the moon is new or full. When the sun, the earth and the moon are at right angles to each other, the combined power of their gravity is weakest and high tides are very low. These are known as neap tides. Neap tides happen when there is a half-moon.

Once every 36 months, the moon will be in line with the sun and very close to the earth. This will produce extremely high tides, which are called proxigean spring tides.

The terms new moon, half moon and full moon are descriptions of the shape of the moon as seen from the earth. The moon we see changes shape over a 29.5-day cycle. These changes are known as the phases of the moon. They are caused by the moon's relationship to the light of the sun. The moon is visible because it reflects light from the sun to the earth. The amount of light the moon is able to reflect is known as its albedo.

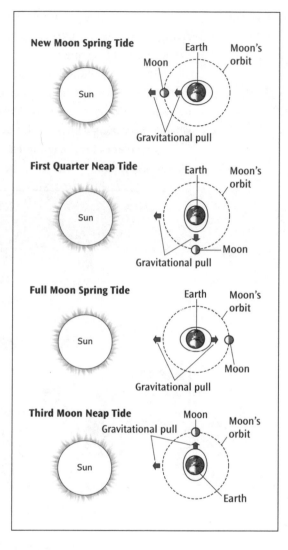

Gravity and the tides.

Albedo is the extent to which any object that does not shine by its own light is able to reflect light. This is measured on a scale of 0–1 where 1 is light and 0 is dark. The moon's surface varies in reflectivity from 0.07 in the darkest parts to 0.24 in the lightest. Its average albedo is 0.12 (the earth's is 0.37).

As the relative positions of the sun, the earth and the moon change so the moon reflects a greater or lesser area of light, and so its shape appears to change.

At the start of the cycle, the moon is invisible to us on earth. It then appears as a slim crescent-shaped sliver with its horns pointing left, and gradually grows into a full bright disc, after which it gets smaller, becoming a crescent facing right, and then disappears.

The names for the complete sequence are shown in the illustration below.

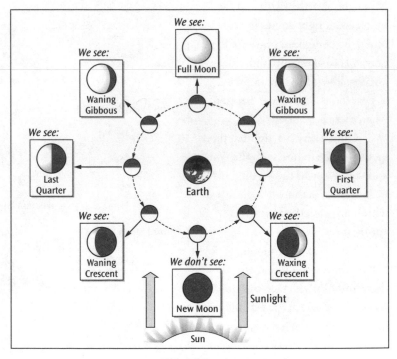

Phases of the moon.

The meaning of waxing, waning and gibbous

When the moon is growing it is said to be *waxing*, a word derived from the Old English 'weaxen', to increase. When shrinking it is said to be *waning*, from the Old English 'wanian', to lessen. And when it is bigger than a semicircle but smaller than a circle the moon is said to be *gibbous*, from the Latin word 'gibbus', meaning a hump.

THE CYCLES OF THE MOON

The moon has many cycles, the most important of which are: the *synodic month*; the *sidereal month*; and the *Metonic cycle*.

Synodic month

The synodic month is the 29.5 days the moon takes to travel from one new moon to the next. In one synodic month, the moon completes slightly more than one orbit of the earth and turns once on its axis. The synodic month and the lunar day are the same thing. The synodic month is also called a lunation.

Sidereal month

The sidereal month is the 27.3 days it takes the moon to make a complete orbit of the earth. The sidereal month is slightly shorter than the synodic month. At the end of the sidereal month the moon will not have rotated once on its axis and it will not have completed its full cycle of phases.

Metonic cycle

The moon appears to wax and wane on a monthly cycle. In fact, it takes 19 years, or 6940 days, for it to reappear in exactly the same part of the sky. This is known as the Metonic cycle. The cycle is named after the fourth-century BC Greek philosopher Meton, who calculated the length of the cycle, although its existence had been known since Babylonian times.

Moon cycle summary		
Synodic	29.5 days	Full moon to full moon (a lunation)
Sidereal	27.3 days	One orbit of the earth
Metonic	19 years	Time to get back to exactly the same place in the sky

APOGEE AND PERIGEE: THE MOON'S DISTANCE FROM THE EARTH

The moon's orbit is elliptical, and its distance from the earth varies. The furthest the moon gets from the earth is 252,716 miles, known as the *apogee*. The nearest it gets is 221,468 miles, the *perigee*.

LIBRATIONS: WHY WE CAN SEE 59 PER CENT OF THE MOON'S SURFACE

As the moon orbits the earth it appears to wobble slightly. This enables us to see more than 50 per cent of its surface.

As the moon orbits the earth it appears to wobble slightly. This enables us to see more than 50 per cent of its surface. It is as though the Man in the Moon were very slightly nodding and shaking his head. As he nods, we glimpse a bit more of the top of his head and under his chin. The same thing happens as his head shakes – we see a bit more of his ears. This process is called libration. There are three forms of libration:

▶ Latitudinal libration: the Man in the Moon nodding, which enables us to see 6.7° more to the north and south.
▶ Longitudinal libration: the moon shaking his head from side to side, enabling us to see 7.7° more to the east and west.
▶ Diurnal libration: the daily phenomenon that allows an observer on the earth to see slightly round the back of the moon's western edge as it is rising and slightly round the back of its eastern edge as it is setting. 'Slightly' means about one degree. Diurnal libration is caused by the motion of the earth relative to the moon.

These three librations enable us to see almost 60 per cent of the moon's entire surface, although never more than 50 per cent at any one time.

ECLIPSES

There are two types of eclipse, *solar* and *lunar*:

▶ Solar eclipses occur when the moon passes between the earth and the sun. Total solar eclipses are quite rare. The last one to be seen in England occurred in 1999, and the next one is not due until 2090.

▶ Lunar eclipses occur when the earth passes between the sun and the moon. They take place only when the moon is full. As the eclipse takes place, the light from the sun is 'bent' as it passes through the earth's atmosphere. This causes the moon to take on a reddish glow.

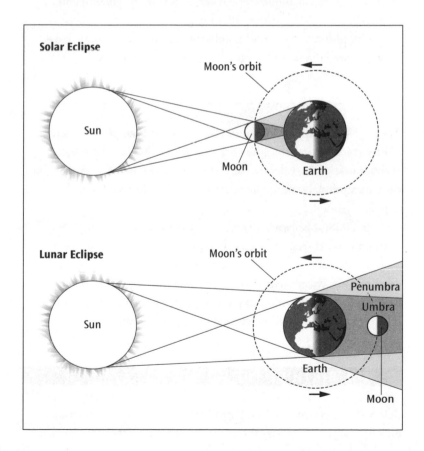

CHEMISTRY OF THE MOON'S SURFACE

All the chemical elements that make up the earth are found on the moon, but in different quantities and configurations. The major elements on the surface are oxygen, iron, aluminium, silicon, magnesium, calcium and sodium. There are also small amounts of titanium. Judged by the abundance of atoms of an element, oxygen is the most abundant element on the moon – 60 per cent of the atoms on the surface are oxygen but are all chemically bound to other elements. The second most abundant element is silicon, at about 15 per cent; then aluminium, at about 10 per cent; followed by calcium and magnesium.

The major elements on the surface are oxygen, iron, aluminium, silicon, magnesium, calcium and sodium.

Continuous meteoroid bombardment and solar winds have contaminated the moon's original chemical composition.

Water

Apollo 15 and Apollo 17 found glass droplets that had been jetted out by volcanic activity. They originated deep within the magma and because of the velocity with which they reached the surface they were exceptionally pure, uncontaminated by the rock through which they had passed.

In 2008, at Brown University in America, analysis using a technique called secondary mass ion spectrometry detected water in the glass droplets at a concentration of 46 parts per million. This suggests that there was water on the moon when it was created and that there may be water frozen somewhere in polar regions, where perpetual shadow maintains the surface temperature at near absolute zero.

THE BIRTH OF THE MOON

Before the Apollo manned landings of 1969 and the recovery of moon rock for analysis, there were three main theories about the moon's origins.

1) Fission theory

Fission is the word used to describe things splitting apart. Fission theory claims that billions of years ago, early in its formation, the earth was a fast-spinning fluid planet. As it spun, it threw off some of its matter. This matter formed a ring around the earth which then combined into a solid body, the moon. It has proved impossible to explain why, if the theory is true, the moon's mantle is richer in iron and refractory elements than the earth's. (Refractory elements are substances like calcium and aluminium that boil at high temperatures.) Another problem with this theory is that viscous rotating fluids, which is what the earth would have been, cannot be spun fast enough to cause fission.

2) Capture theory

The capture theory claims that the moon was formed in another part of the solar system and was captured by the gravitational pull of the earth. The conditions needed to make such a capture possible are improbable. Among other things the theory requires that the moon be formed at the same distance from the sun as the earth and that the earth's atmosphere help slow the moon object. Both are highly unlikely.

3) Co-accretion theory

The co-accretion theory maintains that the earth and the moon were formed at the same time and in the same place from what is called solar nebula (the primeval dust and gas that existed before the formation of the solar and stellar systems). Once formed, the earth and the moon remained gravitationally linked. If this were true, it would be likely that the two bodies would have roughly the same properties – they do not.

After the Apollo landing, a new theory evolved which combines some elements from each of the three classic theories. This is known as the Giant Impact Theory and was first proposed in 1975, by two American astronomers, W. Hartmann and D. R. Davis.

The Giant Impact Theory

Hartmann and Davis suggested that around four and a half billion years ago, the earth was struck by a body about the size of Mars. Most of the mantle and core material of the two planets merged into a reconstituted earth. The residual mantle and a small proportion of the cores were spun off (or jetted) into orbit around the earth. Once in earth orbit the debris fused to form the moon. This theory neatly explains the physical differences between the earth and the moon.

Computer models have been used to illustrate the collision in great detail (see below).

It was assumed that before impact both bodies had a similar composition: an iron core surrounded by a mantle. In the impact, enormous heat was generated and the moon was surrounded by debris that would bombard it for the next 3 billion years.

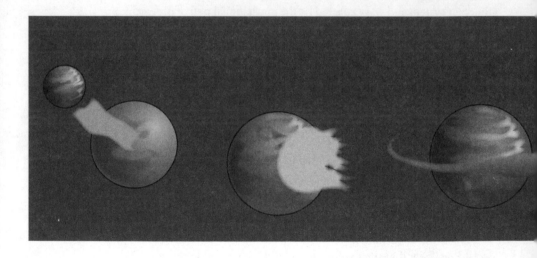

To date, there has been no serious mathematical objection to this theory. It must be remembered, though, that in spite of the major advances made in the light of the data retrieved from the Apollo and Luna missions, we still do not know enough about the moon for the Giant Impact Theory to be anything other than a highly probable hypothesis.

Summary of main theories about the origin of the moon

Theory	Description	Probability
Fission theory	The moon created from material spun off from the earth	Low. Properties of the moon's and the earth's crusts are too different
Capture theory	The moon created away from the earth and trapped into the earth's orbit	Low. Circumstances for this to occur are unlikely
Co-accretion theory	The earth and moon formed at the same time, from the same primeval gas and dust	Low. Properties of the earth's and the moon's crusts are too different
Giant Impact Theory	Mars-size object collides with the earth, and the moon is formed from the resulting debris	High. Explains a lot about properties of the earth and moon

Giant Impact Theory: a Mars-sized object collides with the earth.

Evolution of the Moon

The moon is about four and a half billion years old. At its nativity it was completely or partially molten. The moon's evolution is a history of what happened as the molten material cooled down. Most of that cooling took place in the first 2 billion years of the moon's life. Our understanding of what happened is largely based on data produced by the rock samples retrieved from the manned and unmanned landings on the planet's surface.

THE MOON'S LIFE STORY

The first 100 million years after impact

The earth has been struck by a huge, Mars-sized body. In the next 15 minutes, molten and vaporized matter is jetted from the earth's surface. Most of this material rapidly forms into a molten body orbiting the earth. The moon has been born. The moon travels in a massive sea of debris, the remains of the collision. This debris will bombard the moon's surface for the next 2 billion years.

It is not known if the whole planet was molten at this stage, but it is certain that the surface was covered by an ocean of molten rock – or magma – many hundreds of miles deep. It is cold in space and even a boiling moon cannot fight the chill. The cooling process begins and three things happen: minerals crystallize in the molten magma; a crust begins to form on the surface of the magma, all of which is churned up by rising and descending convection cells. All the while, the surrounding tail of debris bombards the crust and cooling magma, causing destruction and disturbance.

Three types of mineral crystallize in the molten mass to form a variety of ultra-basic rocks:

▶ Olivines. Among the densest materials found today. Rich in magnesium and iron. On earth they dominate the rocks that line the ocean floors.
▶ Pyroxenes. Dense, rich in magnesium, iron and calcium. Found in solidified lava.

▶ Plagioclase anorthosites. Light, more slowly cooled rocks rich in calcium and aluminium. Formed deep in large bodies of molten magma.

The rocks containing olivines and pyroxenes, along with fragments of the crust shattered by debris impact, sink through the churning magma. In the depths of this molten ocean, the lighter anorthosites begin to form and accumulate into 'rockbergs', which then float upwards. The sinking rocks are destined to form a solid mantle. The rising rocks will be found in the moon's crust in the Highlands.

The baby moon has still got 4.4 billion years of growing up to do.

Moon age: 200 million years
The crust begins to harden

As the minerals in the molten magma crystallize into rocks, so the ocean of magma is reduced in size, leaving only pockets of magma beneath the crust. This residue is made up of elements such as potassium (K), rare earth elements (REE) and phosphorus (P) that will not crystallize together. Collectively these are known as kreep. Kreep originally formed below the surface, but in the cataclysms the moon has yet to endure some will be erupted on to the surface.

Moon age: 300 million years
The crust convulses. Volcanic activity starts

At about 300 million years the moon's surface undergoes a convulsion that will last for the next 500 million years. The cooling has left the moon in a very volatile state. The interior temperature is over 1000°C. Dense, hot, light minerals have cooler heavier minerals lying on top of them in an unstable mix. The whole mass begins to turn upside down. The light-density rock from the mantle is thrust up, eventually erupting on to the surface in volcanic activity. Four billion years later, the Apollo missions will recover fragments of these rocks from the moon's Highlands.

This period of convulsion lasts until the moon is about 800 million years old, although the bulk of the activity is over by the time the moon reaches its 400-millionth year. The convulsions leave the moon with a very complex crust. The picture is made more complicated by the fact that in the mantle the strange Kreep elements are melting in the heat produced by their own radioactive decay. They then float up and resolidify in the crust. Some Kreep elements are thrown on to the surface by volcanic activity and will be found in the Apollo rock samples.

Moon age: 800 million years
Cooling nearly over
As single-cell life emerges on earth, the moon enters the last stages of its geological history. The planet becomes much cooler and more stable. Impact marks from the bombarding debris are now preserved rather than destroyed by convulsion. More huge impacts form the giant multi-ringed craters such as Nectaris, Serenitatis and Imbrium. These impacts throw out vast sheets of impacted melted rock containing elements from deep in the crust, mixed with elements from the impacting debris.

Moon age: 2.6 billion years
The moon is almost inert
Another 1.2 billion years have passed since the formation of the huge impact craters. Sporadic volcanic activity has continued throughout the period and lava has flowed into the troughs and basins round the eruptions, forming the enormous dark areas that we call lunar seas or *Maria*. These seas consist of material which is much younger than the basins they fill. They contain minerals that were created at very high temperatures in the mantle, up to 250 miles below the moon's surface.

Analysis of samples taken from the seas illustrates the complexity of the moon's crust and gives scientists more clues about the moon's evolution. Some volcanoes have thrown up material which has crystallized as glass, which suggests to lunar geologists that the moon's deepest interior was never molten.

Moon age: 3.8 billion years
The moon's penultimate giant impact crater, Copernicus, is formed
Lacking any convulsive cooling or volcanic activity to destroy them, the effects of the impacting body can still be clearly seen.

Moon age: 4.4 billion years
The moon's youngest giant crater, Tycho, is formed

Minor impacting continues to the present day, although nobody knows for certain how many objects hit the moon. The seismometers left behind by the Apollo missions recorded 1700 impacts between 1969 and 1972, when they were turned off. It is thought that an object the size of a football hits the moon every day. In 1972, an asteroid the size of a truck and weighing 2400 pounds impacted just north of Mare Nubium. In 2006, a meteorite 10 inches across was filmed as it blasted a crater 45 feet wide and 10 feet deep into the moon's surface. Meteorite strikes could be a problem for long-term inhabitants of any future moon base.

In 2006, a meteorite 10 inches across was filmed as it blasted a crater 45 feet wide and 10 feet deep into the moon's surface.

Moon age: 4.5 billion years
The moon receives its first visitors, Neil Armstrong and Buzz Aldrin

Outline of the moon's evolution	
0–100 million years	Moon formed. Cooling starts. Olivines, pyroxenes and anorthosite-bearing rocks are formed
200 million years	Cooling continues. Kreep elements appear (potassium, rare earth elements, phosphorus)
300 million years	Moon's crust convulses, volcanic activity
800 million years	Huge impact craters Nectaris, Serenitatis and Imbrium formed. Cooling continues and moon stabilizes
2.6 billion years	More volcanic activity, seas formed
3.8 billion years	Copernicus, penultimate giant impact crater, formed
4.4 billion years	Tycho, youngest giant crater, formed
4.5 billion years	Man arrives

Geologists have divided the four and a half billion or so years of the moon's existence into five periods, named after five large craters whose existence spans the life of the moon. The geologists have assumed that the less disturbed a crater is, the younger it is.

The pre-Nectarian period: 4.5–3.9 billion years ago
The first 700 million years of the moon's life
Establishing what happened in the period before Nectaris was formed is guesswork. It is assumed that, as it cooled, the moon's crust was subject to very heavy bombardment, and later in the period to some volcanic activity. Thirty huge impact craters were created in the period, but their features have been almost obliterated since.

The Nectarian period: 3.9–3.8 billion years ago
This short period covers the next 200 million years
The crater Nectaris came into existence 3.9 billion years ago. The basin of Nectaris is very large and very damaged. The period saw the creation of twelve more huge impact basins, and more than 1700 craters that were 12 miles wide or more. Volcanic activity increased as rock within the mantle melted and began to find its way to the surface through the cracks caused by the large impacts. At the end of this period, the moon is only 800 million years old.

Imbrian period: 3.8–3.2 billion years ago
The Imbrian period lasted for 600 million years
It was one of the most active periods on the moon. The impact that created the Imbrium Basin was a major event. Huge cracks were made in the crust. The material thrown out from the craters formed star-shaped rays that dominated a large proportion of the moon's near side. The rays in their turn were destroyed by many smaller impacts. The period has been divided into two parts: Early Imbrian and Late Imbrian. The Early Imbrian period saw a lot of cratering, while the Late Imbrian was the period of greatest volcanic activity.

It was one of the most active periods on the moon.

Eratosthenian period: 3.2–1 billion years ago
The Eratosthenian period lasted over 2 billion years

The Eratosthenes crater is about 60 miles wide. Again, huge star-shaped rays of debris were thrown out and have been largely destroyed by later activity. It was during this period that things began to slow down. Far fewer meteorites hit the moon, and many of the craters became flooded with lava.

Copernican period: 1 billion years ago to the present day

Along with the craters Tycho and Kepler, Copernicus is one of the youngest major lunar landmarks. The rays of material thrown out at its creation are clearly visible: they extend right across the moon's surface and have not suffered much damage. Little else has really happened in the last billion years of the moon's life.

A FILM OF THE MOON'S LIFE

Were we to have made a stop-motion film of the moon's evolution, we might have shot one minute for every 50 million years. This would give us a 90-minute film.

In the first 6 minutes of the film we would see a band of fiery droplets merging together to form a molten, glowing sphere. The surface of the sphere throbs and sparkles as it is bombarded by the debris surrounding it. As it cools, the sphere darkens, accentuating the glittering debris impacts that are now slowing down to about one every 4 seconds. At about 13 minutes, the dark surface convulses with fire as something enormous hits it. This is followed a minute later by a second explosion. The two huge craters Imbrium Basin and Orientale Basin have been formed. Pinprick lights of volcanoes glitter across the surface. Sinister dark lava spreads, and stains the moon's surface. The film has been running for 30 minutes. The moon is 1.5 billion years old. The pace of change slows. Every minute or so there is a new impact. More lava pools form. Action in the last hour of the movie is very slow.

Slowly, the face of the moon takes on its now familiar appearance.

At 75 minutes there is a major disturbance as something big smashes into the surface, forming the crater Copernicus. Six minutes later, there is another big impact and the moon's youngest giant crater, Tycho, appears. The rays of the debris thrown out from the impact stretch for hundreds of miles. The moon is over 4 billion years old.

The moon stares solemnly at the camera, as if trying to understand its strange parent. On earth *Homo sapiens* is beginning to appear. The moon's stare will fascinate the emerging civilizations. They will give it an identity and a will. They will seek its help and fear its anger. It will dominate their magic and their science, and it will send some of them mad. The cold, inert, uncomprehending moon will understand none of this. Eventually, a mankind crazed by two world wars will send a deputation to the moon 'not because it is easy but because it is difficult'. The deputation hopes to enlist the frozen moon's help in a new Cold War. Beneath the moon's impassive gaze Apollo 11 blasts into the sky. When it lands, the next phase of the moon's life will begin: exploration.

The geological ages of the moon

Period	Approx. time ago, billions of years	Moon's age at end of period, years	Geological features
Pre-Nectarian	4.5–3.9	700 million	Moon formed. Crust cools. Surface subject to severe bombardment, creating 30 giant craters. Most since obliterated
Nectarian	3.9–3.8	800 million	Nectaris formed. Some volcanic activity. Heavy bombardment
Imbrian	3.8–3.2	1.4 billion	Imbrium created in giant impact. Many impacts, then extensive volcanic activity
Eratosthenian	3.2–1	3.6 billion	Fewer impacts. Large amounts of lava flow form *Maria*
Copernican	1–present day	4.6 billion	Very little activity. Copernicus and Tycho formed

The Present Structure
of the Moon

It is thought that the moon is made up of three concentric spheres:

▶ An outer low-density crust, up to 90 miles thick.
▶ Beneath the crust a denser mantle about 620 miles thick.
▶ At the centre there may be a small iron-rich metallic core
  100 miles in radius.

Understanding the mantle has been described as the Holy Grail of the
moon. Some of the other important questions yet to be answered are:

▶ How did the lunar crust form?
▶ The moon's crust is thicker on the far side than on the near.
  Why is this?
▶ Does the moon have a metal-rich core?
▶ If so, what is it composed of?

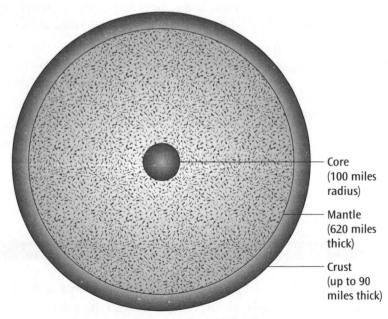

Core
(100 miles
radius)

Mantle
(620 miles
thick)

Crust
(up to 90
miles thick)

The moon's structure.

THE CRUST

The moon's crust is about 40 miles thick on the near side, and about 90 miles thick on the far side. It is composed of rocks rich in aluminium and calcium, which are categorized as a plagioclase feldspar called anorthosite. Some of the surface is covered in lava erupted from the moon's mantle. This is known as lunar basalt. The lunar basalt is rich in iron and magnesium and appears darker than the surrounding rocks of the Highlands which border it. The crust contains oxygen, hydrogen, silicon, magnesium, iron, calcium and aluminium. It also contains some trace elements, including titanium, uranium, thorium and potassium.

THE MANTLE

The mantle lies below the crust and is about 620 miles thick. The mantle is sometimes divided into the lower mantle and the upper mantle, with the division occurring about 300 miles down. The rocks in the upper and lower mantle may be different from each other. The mantle is made up of what is known as mafic material. This is rock composed of minerals rich in iron and magnesium, such as olivine and pyroxene. It was the partial melting of the mantle early in the moon's life which caused volcanic activity. Lava from the mantle has been found to contain titanium. Moonquakes occur deep in the mantle at about 450 miles. They are caused by the gravitational pull of the earth and affect the moon's orbit.

It was the partial melting of the mantle early in the moon's life which caused volcanic activity.

THE CORE

The core (if it exists) is at the very heart of the moon. It is small, possibly about 100 miles in radius, and may be partially molten. It may once have been magnetic.

The Surface of the Moon

The surface of the earth is a very lively place, subject to continual change. No geological forms on the earth's surface have survived from the early formation of the crust. On the moon it is different. Things change very slowly up there. Neil Armstrong's first footprint will be visible in thousands of years. Equipment left on the moon will last for millions of years. It is possible to find rocks on the surface that have survived from the time when the earth and the moon were created. David Scott and James Irwin, from Apollo 15, found a rock that is over 4.5 billion years old – a survivor from the dawn of the solar system. Called the Genesis Rock, it is now in the Johnson Space Center in Houston.

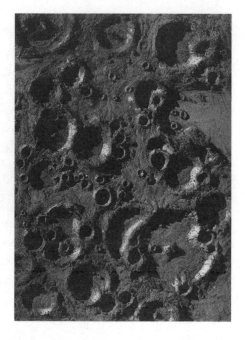

An imaginative nineteenth-century evocation of the moon's surface.

The moon's surface has three main features:

▶ *Terrae*: the light-coloured Highlands.
▶ Seas or *Maria*: the large circular darker areas.
▶ Regolith: a layer of dust that covers almost all the moon's surface.

THE HIGHLANDS (TERRAE)

The Highlands are made up of craters, which are scars left by objects impacting with the moon. Space is full of orbiting rock, the debris left from billions of years of colliding planets and asteroids. These rocks continually bombard the earth, although most burn up in the atmosphere. The moon has no atmosphere to protect it. An object hitting the moon will be travelling at between 10 and 15 miles per second, or around 45,000 miles per hour. The size of the object will range from the very tiny to the very large. A piece of stellar dust will leave a microscopic crater. A stellar rock tens of yards or even miles wide will make a huge scar visible from the earth. The pressure created on the moon's surface at the moment of impact will be anything from 150,000 to 150 million pounds per square inch. This energy is turned into heat, which reaches temperatures of up to 10,000°C – hotter than the sun, and hot enough to cause the rocks involved in the impact to melt, vaporize or even collapse into their constituent atomic parts. This process is known as ionization. The early years on the moon were so violent that many of the original craters have been completely demolished.

Breccias

The heat of impact forms a composite rock known as breccia. Breccias are composed of angular rock and mineral and glass fragments, set in a matrix. The matrix is the substance holding the fragments together and may be formed of smaller particles of the same composition, or different components altogether. Almost the entire surface layer of the moon is made up of breccias.

Crater formation: the first 90 seconds

It is night. A small meteorite, 2 miles wide, travelling at 45,000 miles per hour, is 10 miles above the moon's cold, dark surface. One second later it hits, bringing with it cataclysmic violence. A searing light flashes to the horizon. The impact creates a pressure of 10 million pounds per square inch and the temperature rises towards 10,000°C. The solid rock of the meteorite and the lunar surface around it melt and fuse together. The surrounding rocks become like jelly, losing all their strength.

Vaporized molten material is hurled out of the impact area and shockwaves shudder across the lunar terrain. A bowl-shaped hole only slightly bigger than the asteroid has been blasted into the moon's surface. The unstable floor in the centre of the crater bounces up and then collapses back as the boiling residue of the meteorite and the moon rock fall back on to it. It has been only seconds since impact. The shockwaves rumble on, vibrating through every inch of the surface. Vaporized molten rock falls in slow motion, blanketing everything for miles around. Huge shards of rock weakened by the shock and the heat break off from the rim of the crater and slide into the boiling rock at its base. Slowly all movement ceases. The crater glows white and the tormented rock starts to cool. The moon has a new simple crater. In a couple of billion years, the crater will be photographed, categorized, and given a name.

Not far behind the small meteorite another much larger object is heading towards the moon. This one is 25 miles wide, the size of London. It hits with a force of nearly 45 million pounds per square inch. Again the temperature soars to 10,000°C. In a fraction of a second, rocks in the impact area break down into their atomic parts. The lunar floor turns to jelly and bounces like a drumskin. An enormous curtain of molten rock rains down over hundreds of square miles. At its centre a crater 32 miles wide has been formed. To approach the crater you will have to cross hundreds of miles of debris thrown out by the impact. You will climb two or three thousand feet up the rim, and as you reach the summit you will look down at the walls, broken by terraces, sloping gently down for 7000 feet, deeper than the Grand Canyon. Fifteen miles away in the middle of the vast desolate space you will see the remains of the central peak, which rose and collapsed in the first seconds of impact. Round it there is another rim 5000 feet high, formed by the peak collapsing on to the crater floor as it jellified in the colossal heat. You will be looking at a landscape that, billions of years ago, was created in a few seconds: a landscape that contains clues to the evolution of our solar system.

In 2029, the asteroid Aphophis will come within 19,000 miles of the earth's surface. Aphophis is 300 yards wide and travelling at 12 miles a second. It is thought to have a one in 5000 chance of hitting our planet. The impact would release 1000 megatons of energy, wipe out most forms of life and radically change the earth's climate. For 2 billion years, the moon endured such impacts as a matter of course.

Moon craters are evaluated on a scale of 1–5, where 1 is a fresh undamaged crater and 5 is a battered veteran only just hanging on to crater status.

Types of crater

The moon has three main types of crater:

▶ simple craters
▶ complex craters
▶ basins

Simple craters
Simple craters have a diameter of up to 12 miles. They are usually rounded or bowl-shaped with flat floors. The sides are smooth, but as the diameter of the crater increases, material on the crater wall can break away and slump into the floor of the crater, leaving the rim with a scalloped look.

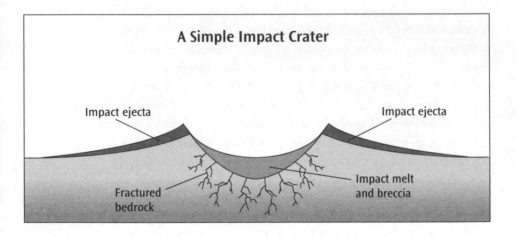

A Simple Impact Crater

Impact ejecta

Impact ejecta

Fractured bedrock

Impact melt and breccia

Simple craters have a diameter of up to 12 miles.

Complex craters
Complex craters have a diameter of between 12 and 90 miles. At the centre of the crater there is a peak, or evidence that there once was a peak. The rims are scalloped and terraced. As the diameter increases to about 50 miles, the floors of complex craters become rougher round the central peaks. As the diameter reaches 60 miles the rough area is replaced by a ring of fragmentary peaks. Larger complex craters may have concentric fragmentary rings of peaks round the central peak.

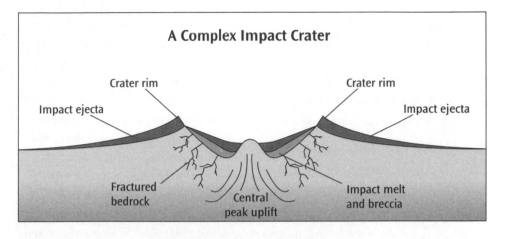

A Complex Impact Crater

Crater rim

Impact ejecta

Crater rim

Impact ejecta

Fractured bedrock

Central peak uplift

Impact melt and breccia

Basins

Basins are the largest form of impact crater on the moon. They range from 90 miles to over 600 miles in diameter. Basins are subdivided into three further sub categories: central peak basins, peak ring basins, and multi-ring basins.

Complex craters range from 12 to more than 90 miles in diameter.

▶ Central peak basins are small basins about 90–100 miles in diameter. They are typified by a fragmentary ring of peaks round the central peak.
▶ Peak ring basins range from 100 to 280 miles in diameter and have a fragmentary ring of peaks, but lack a central peak.
▶ Multi-ring basins are over 250 miles in diameter and are typified by concentric rings round the central peak. They may have as many as six rings. Orientale is the best example of a multi-ring basin.

The debris round the craters

The material thrown out of the craters in the first seconds after impact is called ejecta. The ejecta are thrown out in a curtain of molten rock that lands in star-shaped rays around the impact point. The rays round the moon's youngest crater, Tycho, can be seen extending for hundreds of miles right across the moon's surface. Material from deep in the crater lands nearest to the impact point; this is called continuous deposit – it is coarse-grained and blankets everything. Material from what was the crater's surface lands furthest away; this material is called discontinuous deposit. It is fine-grained and mixes with the material it is landing on. The ejecta form a geological negative of the structure of the crater from which they have come.

Cracks and fissures penetrate right into the 50-mile-thick crust and radiate for hundreds of miles around the crater.

After the violence of impact, the rocks cool to their old hardness. Much damage has been done. Cracks and fissures penetrate right into the 50-mile-thick crust and radiate for hundreds of miles around the crater. This damage has created the openings that will allow the boiling magma coagulating 250 miles underground to find its way to the surface.

THE LUNAR SEAS (MARIA)

Nearly a fifth of the moon's surface is covered in huge, overlapping disc-shaped areas of dark matter. They fill many of the impact basins, the moon's biggest craters. Mankind has always wondered what the grey material is. Early astronomers thought it was water, so named the dark areas 'Seas', in Latin, *Maria*. Apollo 11 landed in the Mare Tranquilitatis, or the Sea of Tranquillity. More recently, astronomers wondered whether the seas were volcanic lava or huge sheets of rock which had melted in the very high temperatures created when the impact basins were formed. Now that it has been possible to examine the rock, it is known that the seas were formed by volcanic activity.

Volcanic activity: what is it?

Put very simply, a volcano is a tube or system of tubes that allows molten material to pass from a planet's interior to its surface. All planets in the solar system are thought to show signs of volcanic activity.

Craters and seas are beautifully imagined in this late-eighteenth-century engraving.

The products of a volcano

Three types of material explode from a volcano:

▶ Lava: magma formed from the melting rock deep in a planet's mantle.
▶ Gas: formed from material which has boiled and evaporated in the heat. These substances are called volatiles and they leave small holes in the lava which are the remains of gas bubbles. The holes are called vesicles.
▶ Solid objects: material ranging in size from minute dust particles to huge rocks, weighing tons.

Volcanic material is important. If it can be examined in the laboratory, it is possible to tell a lot about where and how it was formed. On 24 July 1969, that possibility became a reality when Apollo 11 splashed down into the Pacific Ocean. On board were fifty-eight samples of the rarest commodity on earth: moon rock. Man had not only landed on the moon, but held over 47 pounds of it in his hands, all gathered from the Sea of Tranquillity.

Lava formation

When the young moon was about 300 million years old, decaying radioactive material deep within the mantle began to heat up and melt, forming magma. The moon's surface had been badly damaged by impacts, and by the time the moon was about 700 million years old the molten magma had forced its way through the cracks and fissures to erupt on to the surface as lava. This process went on for another 800 million years until the moon was about 1.5 billion years old. After that, volcanic activity slowed down but continued for the next billion years. Most volcanic activity had stopped by the time the moon was 2.5 billion years old.

The lava that reached the surface had a very low viscosity, and because it was so runny it flowed over vast areas. The giant impact basins were the areas of the moon that were the most damaged and this is where the lava flowed most easily. It covered the basin floors and flowed out through cracks in the rim. The seas can be up to a mile thick and are made of up to three layers of lava from successive eruptions. When one layer of lava flows on to another, the weight can cause the whole structure to bend and crack.

Some of the most interesting volcanic deposits found in the *Maria* were droplets of glass, which are the products of lava fountains. Thin magma, sprayed under great pressure on to the surface, formed a curtain of beads which cooled in flight and landed as glass droplets. Most volcanic material becomes contaminated by the rocks through which it passes on its way to the surface. The glass droplets travelled so quickly that they escaped contamination. They are the purest examples of material from the lunar mantle that have so far been recovered.

They were first identified round the Apollo 15 site, where the droplets were green, and around the Apollo 17 site, where they were black and orange. They have since been found in almost all lunar basalt samples. The surface of the beads shows evidence of lead, zinc and chlorine. Samples also show varying degrees of titanium. It was these droplets that led researchers at Brown University to suspect that there may have been water on the moon at its creation.

Volcanic aftermath

Volcanic activity has left a legacy of strange, beautiful and puzzling land shapes. Some of the most important of these are sinuous rilles, Mare domes, lava terraces and cinder cones. These structures are fifty times larger than similar ones found on earth. This is because the moon's gravity is only one sixth that of the earth, so large rock structures are very light and do not collapse under their own weight.

Sinuous rilles
Rilles are collapsed lava tubes through which molten lava once flowed. They can also be valleys eroded by the flowing lava. It was once thought that they might have been formed by water, but no water or evidence of water has ever been found in them.

Rilles often start on the edges of basins, although some start in the Highlands. The largest concentration is found round the Marius Hills, where there are many volcanic features; and on the Aristarchus Plateau. Rilles are narrow winding valleys with V- or U-shaped floors. They all run downhill and disappear into the *Maria*. They can be nearly 2 miles wide, half a mile deep and up to 180 miles long. The largest is the Hadley Rille in the Mare Imbrium.

Mare domes
These are low, broad, convex, circular or oval landforms, with a ground diameter of between 1.2 and 15 miles. They range in height from 300 to 600 feet and have roofs which slope up gently at 2–3 degrees. Eighty of these domes have been mapped on the surface of the moon, and most are in the Marius Hills.

Lava terraces
These resemble a shoreline and are found in basin rims and at the boundaries with the Highlands. They may be evidence of a lava tide which ebbed away while still molten, possibly flowing back into the fissure through which it emerged.

Cinder cones

These are the classic volcano shapes. On the moon they are found in association with rilles. They are less than half a mile high and 1–2 miles wide at the base. Some appear to have craters in the summit. They are sometimes found in linked chains, and these are thought to be fissures.

Present-day volcanic activity and transient lunar phenomena

It has been generally assumed that the moon is volcanically inert and that it has been that way for the last 2 billion years. However, over the centuries there have been reports of bright lights and curious goings-on on the moon's surface. In the tenth century monks at Canterbury cathedral reported:

> *There was a bright new moon, and as usual in that phase its horns were tilted towards the east; and suddenly the upper horn split in two. From the midpoint of this division a flaming torch sprang up, spewing out, over a considerable distance, fire, hot coals, and sparks.*

Since then there have been many such reports of similar events. They are called transient lunar phenomena (TLP). Witnesses include the British astronomer Sir William Herschel, in the eighteenth century, and the crew of Apollo 11. Many of the sightings are from amateur astronomers and are difficult to assess. It is thought that the sightings are the result of 'outgassing' – gas building up under pressure within the moon and exploding through the regolith. A recent theory is that the moon still undergoes volcanic activity. A systematic attempt is being made to record TLPs. A 10-inch robotic telescope at Cerro Tololo in Chile is being used to check for TLP activity and to alert astronomers round the world so that data can be collected and analysed.

It is possible that as the moon expands and contracts under the influence of the earth's gravity, lunar rocks are ground together and produce gas. Future manned and unmanned landings will release

30 tonnes of gas on to the surface of the moon with each landing. This will contaminate the atmosphere and make analysis of natural outgassing phenomena impossible.

THE REGOLITH

For more than 4 billion years, the moon has been bombarded by meteoroid objects ranging in size from tiny particles of dust to enormous objects the size of Manhattan. Like hellfire this has churned and rechurned the surface. Rocks have been ploughed up, buried, reploughed, and reburied for billions of years. Major meteor bombardment stopped more than 3.5 billion years ago, but for all its life the moon has been scoured by small particles travelling at 33,000 miles per hour. Something the size of a small truck impacts every hundred years or so. Under this onslaught, the surface rock has been reduced to a fine dust. This is called the regolith.

Major meteor bombardment stopped more than 3.5 billion years ago, but for all its life the moon has been scoured by small particles travelling at 33,000 miles per hour.

The older the rocks on the surface, the longer they have been exposed to bombardment, and the deeper the regolith. In the Highlands, which are the oldest, most exposed places on the moon, the regolith is 20–30 metres deep. On the cooled lava which forms the lunar seas, the regolith is 2–8 metres deep. On the melted rock which forms the floor of the youngest craters such as Tycho, the regolith is sometimes only tens of centimetres deep.

Under the regolith it is thought that there is much coarser debris, called the megaregolith, that was created in the moon's youth. Very little is known about the megaregolith.

The regolith contains many stories. It is the source of almost all our knowledge of the moon. No rock sample has been taken from below the regolith. Trapped in it are rays from the sun and cosmic ray particles from beyond the solar system. If we can disentangle the clues, the regolith will contain some of the keys to the history of the cosmos: the sun, the earth and, of course, the moon itself.

Other Geological Forces at Work on the Moon

After volcanic and impact activity there are two smaller forces at work shaping the moon's surface: tectonic activity and moonquakes.

TECTONIC ACTIVITY

Tectonic activity is a phrase which describes what happens when the surface of a planet is changed by forces which squeeze or stretch it. The process results in a cracking or wrinkling of the surface. If the pressure is applied over a long period and the results appear slowly, the phenomenon is known as *gradual* or *plastic* rupture. If the pressure is sudden and the results appear abruptly, the phenomenon is called rupture.

Tectonic activity occurs in impact basins into which lava has flowed. The weight of the lava causes the crater floor to sag and buckle. The rocks are squeezed to the centre of the crater and stretched at the edge. The visible evidence of this is a wrinkling of the rock at the crater's centre, and cracks or faults at its edge. The best examples are to be seen around the Humorum and Serenitatis basins.

MOONQUAKES

The moon is subject to minor seismic activity. There is no sound on the moon's surface because the atmosphere is too thin to carry sound vibrations, but the noise from seismic activity travels for a very long way through the hard rock. An Apollo geologist said, 'The moon rings like a bell.'

There are three types of lunar quake:

▶ *Deep-focus moonquakes* happen on a monthly basis and may be caused by earth–moon tidal stresses. They occur about 450 miles beneath the moon's surface.

- *Thermal moonquakes* are caused by the heating of the surface at the end of the extreme cold inflicted by the lunar night.
- *Shallow moonquakes* may be caused by tectonic activity and collapsing rock formations. Shallow moonquakes can reach 5.5 on the Richter scale.

The Moon's Future

The moon is a very stable, quiet environment. It is only three days' rocket trip from the earth. If the inherent technical problems can be solved, it could have several uses. It could become an observatory, a launch pad for future deep-space missions or an open-cast mine.

THE MOON AS AN OBSERVATORY

The moon is probably one of the best platforms in the universe on which to mount astronomical instruments. Optical telescopes could be mounted in series, which would give lenses with apertures equivalent to kilometres across. Craters could be turned into housings for radio antennae. The lack of radio clutter and the absence of an ionosphere would allow astronomical investigation across every wavelength, from high to low.

THE MOON AS A LAUNCH PAD FOR DEEP-SPACE EXPLORATION

If bases, as well as construction and manufacturing facilities, could be established on the moon, deep-space exploration would become much easier. The technical problems are enormous. To exploit the moon's potential as a launch pad, we first have to work out how to live there on a permanent basis. Then we would have to find a way to manufacture and launch space vehicles in a lunar environment.

The moon has, among other things, oxygen, hydrogen, aluminium, iron and the isotope Helium-3. It is now thought it may even have water. All of these could be extracted from the lunar rock – a very expensive process that could be made cheaper if energy and building materials to service the moon base could be sourced locally. The regolith could provide other construction materials, which might include a glass strong enough to build with and pure enough to manufacture the finest lenses ever made.

The Holy Grail is Helium-3. This occurs in the first metre of the regolith, which covers more than 90 per cent of the moon's surface. Helium-3 could be used on earth as a source of power in non-radioactive nuclear fusion reactors. Such reactors do not yet exist, but, when they do, they will be economical and very safe – far safer than those using conventional nuclear power. It would be feasible to build Helium-3 power stations very close to major cities without danger or the risk of pollution. Twenty-five tons of Helium-3, or a shuttle-load, could provide the United States with enough energy for a year. One ton of Helium-3 would be worth $4 billion. Practically and financially, mining the moon is an exciting and beguiling prospect, but there would be a cost.

The extraction of one ton of Helium-3 will consume a million tons of regolith, which has to be heated to a temperature of about 800°C. Digging the quarries, and building the living quarters, the observatories, the smelting plant and the factories to service the mines, will destroy everything that makes the moon unique. The surface will become a vast frontier construction site surrounded by millions of acres of torn land and billions of tons of waste soil. Where there was once a pristine near vacuum there will hover a deadly cloud of toxic gas. Trillions of dollars and huge amounts of human resources will be tied up in a lunar Klondike at a time when 90 per cent of the people on earth do not have a simple source of clean water. The future will ask us: did we know what we were doing or were we drunk on moonshine?

Some interesting lunar craters

Crater	Diameter, miles	Description
Albategnius	85	Crater sketched by Galileo, situated in central Highlands
Alphonsus	74	Old, with three volcanic cinder cones on its floor
Archimedes	51	Floor flooded by lunar basalt
Aristarchus	25	New crater
Compton	100	Central peak plus ring basin
Copernicus	58	Young large crater. Copernican period named after it
Descartes	30	Old crater in central Highlands near Apollo 16 landing site
Eratosthenes	36	Eratosthenian period named after it
Flamsteed	70	An old crater with flooding from very recent lava (about a billion years ago). Surveyor 1 landed here
Frau Mauro	60	Landing site of Apollo 14. A very old crater
Herigonius	9	Near some very good sinuous rilles
Kopf	26	A crater which was thought to have been volcanic in origin
Lichtenberg	12	Crater with rays and covered in very young lava less than a billion years old
Linné	1	Tiny crater that was thought to appear and disappear
Shorty	68	Near Apollo 17 landing site
Theophilus	62	Large crater at the edge of Mare Nectaris
Tycho	53	New rayed crater whose rays extend across the whole face of the moon

Lunar *Maria*

Name	Age, years	Description
Mare Crisium	About 3.5 billion	Has a concentration of mascons
Mare Fecunditatis	3.4 billion	
Mare Humorum	3.2–3.5 billion	Contains mascons
Mare Imbrium	Less than 2–3.3 billion	Contains mascons
Mare Nectaris	3.5–3.8 billion	
Mare Nubium	3 billion	
Mare Serenitatis	3.3–3.8 billion	Contains mascons
Mare Smythii	1–1.5 billion. Very young	Contains mascons
Mare Tranquilitatis	3.8 billion	Site of Apollo 11 landing
Oceanus Procellarum	Less than 1 billion	Contains very young lava

Lunar basins

Name	Description	Diameter, miles
Crisium Basin	Nectarian period	460
Humorum Basin	Nectarian period	500
Imbrium Basin	Imbrium period named after it.	720
Nectaris Basin	Nectarian period named after it. Very old (more than 3.8 billion years)	530
Orientale Basin	Youngest multi-ring basin on moon, formed 3.8 billion years ago. Used as a guide to interpret older basins that have been destroyed	580
Procellarum Basin	Largest sea on the moon. It covers several impact basins, one of which may be the 'Procellarum Basin', though evidence for such a basin is sketchy.	2000

Schrödinger Basin	Formed after Imbrium but before Orientale	200
Serenitatis Basin	Site of Apollo 17 exploration, about 3.7 billion years old	560
South Pole – Aitken Basin	Oldest basin on the moon and the largest, deepest impact crater in the solar system. Possibly more than 4.3 billion years old	1500

Other interesting lunar features

Feature	Description
Apennine Bench	Area includes kreep volcanic plains
Apennine Mountains	Large mountain chain on rim of Imbrium Basin
Hadley–Apennines	Informal name given to region of Apollo 15 exploration
Marius Hills	Could be a large volcano. Contains many small domes, cones and sinuous rilles
Reiner Gamma	Bright deposit in Oceanus Procellarum of unknown origin
Rima Hadley	Long sinuous rille, possibly a lava tube or channel
Taurus–Littrow	Informal name given to Apollo 17 exploration area. Has many geological features including *Maria*, dark mantle and Highlands
Tranquillity Base	Site of first lunar landing

Chapter 2
Astronomers

From the moment Neil Armstrong's foot touched the moon's surface, we have been able to examine it in greater and greater detail. We now know a lot about the moon's age, its atmosphere and its construction. We have subjected moon rock to microscopic examination in the laboratory. We have left mirrors on the moon's surface that bounce back the light from our lasers supplying detailed data on the moon's movements. We can discuss the moon's atmosphere in terms of parts per million, and measure seismic activity from deep in the moon's mantle.

It has been a long haul. For thousands of years astronomers had no telescopes and relied on their eyes. Recorded astronomy began in Mesopotamia about 1800 years before the birth of Christ. The astronomers used instruments that would not fundamentally change for 3000 years. Those early detailed observations were recorded on clay tablets in a script known as 'cuneiform'. Technical advances were painfully slow, parchment replaced clay and paper replaced parchment. The measuring equipment got bigger and more accurate.

Recorded astronomy began in Mesopotamia about 1800 years before the birth of Christ.

Before written records our ancestors enshrined their findings about the sun and moon in huge stone structures known to us as megaliths. Even further back in time our Stone Age ancestors, equally obsessed with the moon, recorded its movements on fragments of bone and stone.

In the 1970s the fibula of a baboon was found in a cave known as the Place of Heaven, in the Lebombo Mountains in Swaziland. The bone was 37,000 years old and had twenty-nine notches carved in its side. It is believed that the notches refer to the phases of the moon, which repeat over a 29-day period.

Another bone was found in another cave in the Dordogne Valley in France. It came from an eagle's wing and was over 30,000 years old. Carved on its side were a series of dots, which also describe the phases of the moon.

Those early moon watchers were taking the first steps on the same road that the NASA scientists still tread. Crouched over a bone in a cave notching the phases of the moon or crouched over a computer in a laboratory measuring refracted moonlight, the intention is the same: how to record and understand what is happening up there.

Astronomy can be crudely divided into four overlapping phases:

▶ 35,000–4000 BC: before written records.
▶ 4000 BC–AD 1609: written records, pre-telescope.
▶ 1609–20th century: evolution of the telescope and advanced electronic observation equipment.
▶ 1969–21st century: interplanetary exploration.

35,000–4000 BC
Before written records

Not much is known about the first 30,000 years or so of moon watching. The bone from the eagle's wing is one of just a few clues that are testament to more than 25,000 years of moon watching. The fragments that survive would scarcely fill a large packing case and what is known for certain about them could be written in a small notebook. They all come from a time when mankind had to hunt to survive.

Some of the most important objects are described below.

LEBOMBO BONE
APPROXIMATE DATE 35,000 BC

The Lebombo Bone was discovered in a cave in the Lebombo Mountains in Swaziland. It is made from the fibula of a baboon. The bone has twenty-nine notches carved on its shank and may be the earliest known mathematical calculator or lunar calendar. Similar sticks are still used by the Bushmen in Namibia and are known to have been used in the nineteenth century by the Winnebago tribe in North America.

BLANCHARD BONE
APPROXIMATE DATE 32,000 BC

The Blanchard Bone comes from an eagle's wing. On it are carved twenty-nine notches that can be interpreted as describing the phases of the moon.

ISTURITZ BONE
APPROXIMATE DATE 25,000 BC

The markings on this bone could represent a five- or four-month lunar calendar.

ISHANGO BONE
APPROXIMATE DATE 20,000 BC

A small animal bone that was discovered in the African fishing village of Ishango on the border of Democratic Republic of Congo and Uganda. The bone is inscribed with notches that, like the Blanchard Bone's markings, can be interpreted as describing the phases of the moon.

LASCAUX CAVE PAINTINGS
APPROXIMATE DATE 14,000 BC

Among the most impressive records left by our hunter-gatherer ancestors are the cave paintings at Lascaux in the Dordogne Valley, France. These pictures were discovered in 1940 and show hunting scenes. Among the images are a horse with twenty-nine dots which could represent a lunar cycle and a deer with a line of thirteen dots and an empty square that could represent half a lunar cycle with the empty square standing for the invisible new moon.

Among the images are a horse with twenty-nine dots which could represent a lunar cycle

MEGALITHS

The last phase of the prehistory of astronomy started over 7000 years ago when we began to farm to survive. As we settled into fixed communities we began to erect enormous edifices now known as megaliths.

A megalith is a large stone that forms or is part of a prehistoric monument. Megaliths can be found all over the world, but there is a preponderance of them in northern Europe, especially France, the United Kingdom and Ireland. Exactly why they were built is not known, although there is strong evidence that one of their functions was to act as huge permanent predictors and markers of the orbits of the sun and moon. On the journey from hunter-gatherer to farmer an understanding of the passage of time and of the evolution of the seasons, and an idea of what might happen next, would have been crucial. Astronomical data would have taken hundreds of years to collect and without a method of recording them would have been very vulnerable. This may explain why megaliths are so big. Once built the data they contained was preserved in an immovable and almost indestructible form. The modern dilemma is that the keys to reading the data have been lost.

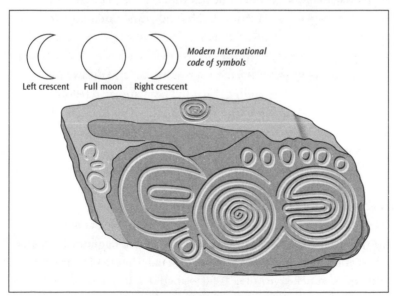

Left crescent   Full moon   Right crescent

Modern International code of symbols

Lunar inscription on stone – Newgrange, Ireland.

We do not know what combination of religious ceremony, fear and scientific curiosity caused our ancestors to spend hundreds of years and millions of man-hours building the megaliths but we can empathize. We still do it. We have spent billions of dollars on, and dedicated millions of man-hours to, mapping and exploring the moon. Nothing we have discovered has any bearing on the lives of almost the entire population of the planet but that hasn't stopped us.

Lunar symbols found on megaliths

In order to analyse the motion of the moon relative to the sun and the earth it is necessary to have a fixed, permanent observation point, a way of counting and a system of notation. All these were available to Neolithic societies.

The most common moon symbols carved into the stones of the megaliths are crescents, circles and wavy lines. The crescents and circles are simple representations of what the moon looks like as it passes through its phases from new to full to old. The wavy line is a visual description of the moon's course as it meanders above and below the path of the sun

Once we understood and could draw how the moon moved we had a way to measure time. The wavy line seems to be an engraved record of the passage of days, months and years, extending back into the past and forward into the future. The new knowledge enabled us to count the days of our lives by the moons we had lived through. Perhaps, as we watched the moon grow, wither and die, we took comfort in the understanding that what has happened before will probably happen again. In the frozen depths of winter spring was waiting. Tomorrow the crescent moon will appear in the empty sky. The stones guarantee it.

Some of the most important megaliths with lunar connections are:

Nabta Playa, Abu Simbel, Egypt
At Nabta Playa near Abu Simbel in Egypt is the site of the earliest known megalith. It dates from before the fifth millennium BC. Nabta Playa is situated in the desert about 500 miles south of Cairo. It is similar to Stonehenge but is 1000 years older. It is believed to be one

of the earliest archeo-astronomical devices in the world. Its precise function is not understood. There is evidence of sacrificial cattle in the area which could be an early manifestation of the Ancient Egyptian Hathor Cult. Hathor was a goddess who protected the night. In some versions of Egyptian mythology she is the wife of the moon god Thoth.

Boyne Valley, Ireland

The Boyne Valley is the site of three enormous mounds. They are over 5000 years old and were originally covered in quartz. They predate the pyramids. The mounds are called Newgrange, Knowth and Dowth. They appear to have been laid out in a system which reflects the movements of the sun and moon. Many of the stones in the three mounds are covered in detailed symbols. Many of the symbols describe the sun. Some of them refer to the moon.

The mound known as Knowth has special lunar associations. It is the site of a stone called the Calendar Stone, which has been described as a megalithic computer. The inscriptions on the stone could be a lunar/solar calculator. The symbols refer to the sun and the moon and describe the 19-year period which the moon needs to reappear in precisely the same part of the sky. A simpler stone bears similar but less-complicated symbols and is thought to be a lunar calculator.

Calendar Stone – Knowth, Ireland.

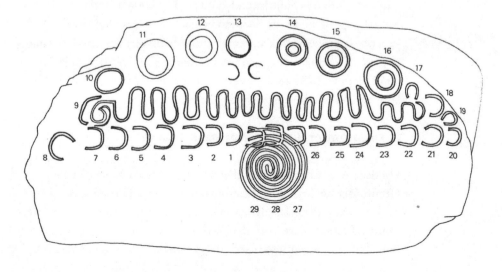

Stonehenge, England

The construction of Stonehenge began about 3100 BC. Recent archeo-astronomical evidence has shown that some of the stones are aligned to mark the extreme position of moonrise in the south and moonset in the north.

Stonehenge.

Callanish, Isle of Lewis, Scotland

This megalith was built around 2900 BC. It is known as the Stonehenge of the north. Some archeo-astronomers think that it is part of a complex of linked monuments which were designed to frame important moonrise and moonset positions. This idea has caused a lot of argument. However, in certain years the midsummer full moon appears to skim the horizon in a ghostly and prophetic way.

Locmariaquer, Carnac, Northern France

On the site at Locmariaquer there is a fallen stone known as Le Grand Menhir Brisé (the great broken menhir). The stone is broken into four pieces. It was once 20 metres high and weighs 280 tonnes. Its sides have been flattened by ancient masons and its tip is polished. It was put into place 1000 years before the much smaller

(25 tonne) stones of Stonehenge. It stood at the centre of a vast complex that was over 12 miles wide. Most of the complex is now submerged under the Golfe de Morbihan. The complex is aligned to mark the extremes of moonrise and moonset in the summer and the winter. Hundreds of years were needed to collect the data to erect the stones of Carnac. The people who built these monuments may have thought they were marking the spiritual and physical centre of the universe, a megalithic Mecca.

4000 BC–AD 1609
Written records, pre-telescope

By 1500 BC, the peoples of Mesopotamia had started to make written records of what they could observe in the sky. The people who made the measurements were astrologers who wanted to predict and interpret eclipses and other important events. The astrologers were part scribes, part scientists and part priests. Even by the standards of modern astronomy their work was very accurate. The technology they used did not really change for over 3000 years. The instruments that Ptolemy used 100 years after the birth of Christ were very similar to those used 1700 years later by the last great pre-telescope astronomer, Tycho Brahe.

SOME OF THE INSTRUMENTS USED BY ASTRONOMERS BEFORE THE TELESCOPE WAS INVENTED

Armillary sphere

'Armilla' is the Latin for bracelet or circle and the armillary sphere is a series of metal circles arranged in varying degrees of complexity to describe the apparent motions of the stars around the earth.

The armillary sphere was adapted by Muslim astronomers and became the astrolabe. Smaller and handier than the armillary sphere, the astrolabe could perform more than a thousand measurements.

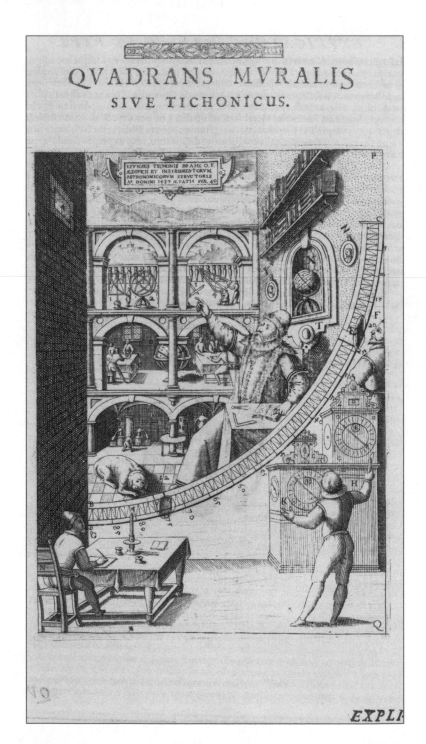

*Tycho Brahe
with his giant
wall quadrant.*

Quadrant

The armillary sphere was complemented by the quadrant. Quadrants were easier to use and could make a large number of complex measurements, including the altitudes of celestial objects, the distances between them and the time of day.

The bigger the quadrant, the more accurately it could be calibrated, and there are some very large examples of them. At the age of twenty-two Tycho Brahe designed a 19-foot quadrant which was so heavy it took twenty men to erect it. An even bigger quadrant from the eighteenth century survives in India at the observatory of Janta Mantar in Jaipur.

Sextant

This instrument was used to measure the position of stars. Sextants came in two types, mobile and fixed. The Islamic astronomers specialized in huge wall-mounted sextants. The great astronomer Ulugh Beg used the largest fixed sextant in the world. It was constructed as part of a wall and had a radius of 40 metres.

Sextants were tools capable of making very accurate measurements, but they could not probe, penetrate or magnify. Like Stonehenge, they were tied to the surface of the earth.

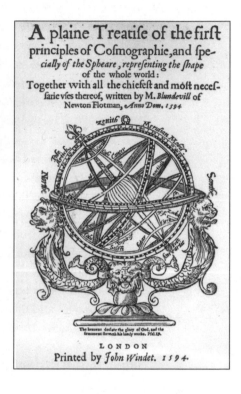

An armillary sphere used to represent circles including those in the heavens, equator and tropics.

MESOPOTAMIAN ASTRONOMY

Mesopotamia means 'the kingdom between two rivers'. It is one of the most important places in the history of the Western world and lies in the area between the Euphrates and the Tigris, roughly where modern

Iraq is now. The tribes that flourished there for nearly 5000 years are sometimes loosely referred to as Babylonian. The Babylonians were the first people to write down what they observed about the moon in a form that we can read. Babylonian culture has had a profound influence on Western civilization. Babylonian writing is known as cuneiform – a series of symbols stamped into clay tablets. It probably started as a way for merchants and farmers to record transactions and quantities, and developed into a writing of great sophistication. Some of these tablets have survived to the present day.

The earliest cuneiform astronomical writings date from about 1800 BC. They record the times of moonrise, the dates of new moons and observations about the weather. The most-comprehensive records are contained in two clay tablets known as Mul.Apin. The tablets are catalogues of the stars and constellations and a record of the constellations that the moon passes in front of. Babylonian astronomy reached its peak between 300 and 75 BC. Over 300 clay tablets survive from this period. They contain a wealth of astronomical data, much of it relating to the moon. Alongside the tables are instructions as to how they are to be used. Data collected and recorded by the Babylonians would be used by the ancient Greeks, and their work would in turn be used by European astronomers some 1500 years later.

Important Babylonian astronomers

Fifth century BC
Nabu-Rimanni (sometimes known as Naburianos or Naburiannu)
Nabu-Rimanni is the first Babylonian astronomer whose name we know. He devised a system of tables for calculating the positions of the sun, moon and stars. The tables are known to us as System A.

Fourth century BC
Kidynas (sometimes known as Kidinnu, Kidunnu, Cidenas, Kidenas)
Several references to an astronomer of this name can be found in Babylonian, Greek and Roman texts, though it is not certain that they all refer to the same man. Kidynas was head of an astronomical

school at Sippar, and he is credited with having made some remarkable discoveries. He is thought to have discovered that the slow rotation of the earth on its axis results in a slight variation in the length of the year. Drawing on the work of Nabu-Rimanni, he improved the tables for calculating the motion of the sun, moon and stars. The tables survive with the name System B. Kidynas is thought to be the astronomer who divided the year up into twelve lunar months. This system is still in use in the Jewish calendar. Incredibly, Kidynas calculated a lunar month to within one second of its true time. His work is thought to have been used by Ptolemy. A damaged clay tablet from Babylon states that 'Ki-du-nu was killed by the sword', and gives the date as 14 August 330 BC.

Third century BC
Sudines
Sudines lived with the Greeks. It was Sudines' tables that the Greeks used to compute planetary motion. They remained in use for nearly 200 years until they were superseded by the work of Hipparchus and then Ptolemy. Sudines is credited with starting the practice of giving astrological meaning to gemstones. This has developed into a complicated pseudoscience.

EGYPT

It is indisputable that the Egyptians built a great civilization, but they were not great astronomers. The moon was very important to them and was associated with some of their most important gods, such as Thoth (see the chapter on 'Gods and Myths'). But the true inheritors of Babylonian astronomy were the Greeks. They took the mathematical formulae and records devised in Mesopotamia and turned them into three-dimensional models.

THE ANCIENT GREEKS AND ROMANS

The Greeks used observation, long-term records and common sense to explain the universe. They believed that mathematics was at the heart of the cosmos. The astral movements that interested the Greeks were the same as those that interested the Babylonians:

▶ The rapid east–west motion of everything in the sky over a 24-hour period.
▶ The slower west–east motion of the sun, the moon and the planets against the constellations.
▶ The repeating short-term moments when the planets appeared to be going backwards.

Pythagoras (*c.*580 BC) led the first school of thought to develop a theory of astronomy. He knew that it was possible for the sun to be at the centre of the universe, but he was persuaded to give this position to the earth. A universe with the earth at the centre is known as a geocentric universe. The idea that the universe was a series of concentric crystal spheres was thought up by Eudoxus (408–347 BC), visualized twenty-seven concentric spheres with the sun, the moon and the stars each having their own separate sphere. This conception of the universe as a series of concentric spheres with the earth at its centre became the orthodox theory for the next 2000 years in the West. The astronomer Aristarchus of Samos (310–230 BC), took the opposite view. Aristarchus is sometimes referred to as the Greek Copernicus. He devised theories about the properties of the sun and placed it at the centre of the universe. Nearly 2000 years would pass before he was taken seriously.

The two most influential Greek astronomers were Hipparchus (mid second century BC) and Ptolemy (AD 85–165). Hipparchus has been described as the greatest astronomer of all time. Although his original works have been lost, they survive because of Ptolemy, a Greek citizen living in the Roman Empire who was heavily influenced by Hipparchus. Ptolemy's great work is called *Syntaxis Mathematica*, or *The Mathematical Compilation*, which came to be known as *The Great Compilation*. In thirteen volumes it describes everything that was then understood about the universe. Volumes 4 and 5 are entirely devoted to an analysis of the moon. *The Great Compilation* survived the fall of the Roman empire. In the ninth century AD it was translated by Arab astronomers as

Almagest (*The Great Book*). The Arabic was translated back into Latin and Greek and appeared in Europe in 1175. The book had a huge impact. Ptolemy's great work and misguided conclusions dominated Western astronomical thinking for the first four centuries.

In 1543, the astronomer Copernicus lay dying. Into his hands had just been laid the first printed copy of his life's work, *De Revolutionibus Orbium Celestium* (*On the Revolutions of the Celestial Spheres*), in which he argues for a universe with the sun at the centre. His theory took some time to make its way in the world and caused much trouble, but in the end everybody agreed that the universe is heliocentric; Aristarchus had been right all along.

Pythagoras led the first school of thought to develop a theory of astronomy. He knew that it was possible for the sun to be at the centre of the universe, but he was persuaded to give this position to the earth.

Important Greek astronomers

Anaximander *c.* 610 BC
Anaximander believed that the moon shone by reflected light and that the earth turned on its axis.

Thales (625–546 BC)
Thales postulated that the earth was flat and surrounded by water.

Pythagoras (*c.*580 BC)
Pythagoras put a fire, which was not necessarily the sun, at the centre of the universe, but he came to believe that the universe was geocentric. He argued that the distances of the planets corresponded to the intervals of the scale in music.

Anaxagoras (500–428 BC)
Anaxagoras explained the nature of lunar and solar eclipses and conjectured that the moon had hills and valleys like the earth. He thought the world to be cylindrical and that the stars were attached to a sphere which revolved around that cylinder. He believed that the

moon shone because it reflected light from the sun and that eclipses happened when the shadow of the earth fell on the moon. He also believed that the earth was located between the sun and the moon.

Meton (c.423 BC)

Meton discovered and defined the Metonic cycle, which states that the moon will come back to exactly the same point in the sky only once every 19 years.

Eudoxus (408–347 BC)

Eudoxus devised a scheme of the cosmos which put the earth at the centre of the universe. The sun, the moon and the planets were contained in twenty-seven geocentric spheres.

Callipus (c.370 BC)

Callipus detected a quarter-day error in Meton's theory. He quadrupled the Metonic cycle and took away one day. The Callipic cycle is 19 × 4 = 76 years minus one day.

Aristarchus (310–230 BC)

Aristarchus' only surviving work is *On the Size and Distance of the Sun and Moon*. He calculated that the sun was nineteen times further away than the moon and that the sun's diameter was twenty times that of the moon. His calculations were right, but his conclusions were wrong. Aristarchus' problem was that he did not have sufficiently accurate measuring instruments. His most important theory was that the universe is heliocentric, that the sun stood at its centre. Aristarchus' ideas were rejected in favour of Ptolemy's, who argued that the universe is geocentric with the earth at its centre. Nearly 2000 years later Aristarchus won the day. His ideas were revised and reworked by first Copernicus and then Johannes Kepler and Isaac Newton. And, of course, we now know that the sun is at the centre not of the universe but of our solar system.

Hipparchus (mid second century BC)

Hipparchus was the greatest observer of antiquity and possibly of all time. He was the founder of scientific astronomy and constructed solar and lunar tables which allowed the positions of the sun and moon to be accurately found, and this permitted the prediction of solar and lunar eclipses. His works are lost but are known to us through his successor, Ptolemy.

Ptolemy (AD 85–165)

Ptolemy was an Egyptian Greek astronomer and mathematician. His great work is *The Mathematical Compilation*. He also wrote *Tetrabiblos*, a work on the then respectable field of astrology.

Ptolemy's *Mathematical Compilation* would be the standard work on astronomy for the next 1500 years. Rome collapsed and Europe entered the Dark Ages. The flame of astronomy flickered. It was saved by the culture of Islam.

ISLAMIC/ARABIC ASTRONOMY

The Mohammedan philosophy of Islam – the Brotherhood of all the Faithful – established an empire that stretched from Morocco on the coast of the Atlantic, across Europe and India to the steppes of mid-Asia. Like the Babylonians 3000 years before them, the Arabs were keen observers and recorders of information.

Islamic astronomers translated the work of the ancients, especially Ptolemy. His *Mathematical Compilation* was called by them *Almagest – The Great Book*. It is still known as *Almagest*. Many Islamic astronomical terms have been incorporated into modern astronomical usage – including azimuth, zenith and nadir.

From about AD 700 to 1200 the Arabs built great observatories and designed and refined astronomical instruments. These include the Baghdad observatory built in the early years of the eighth century and the impressive geared mechanical astrolabe designed by Ibn Samh in 1020. The most important contribution of Arab astronomers was the translation of Ptolemy's work, and the development of an improved theory of the moon's movements.

Some important Islamic astronomers

Albategnius (Al-Battani, Muhammad bin Jabir) (858–929)

Albategnius was a very important astronomer. In a forty-year period, he made detailed tables of the motions of the sun and the moon. His work was published as *Kitab al Zij* and was translated and summarized in the Middle Ages as *De Motu Stellarum* (*On the Motion of the Stars*). It was one of the major books of the sixteenth and seventeenth centuries and influenced many astronomers, including Tycho Brahe, Kepler, Galileo, Copernicus and Hevelius. The lunar crater Albategnius is named after him.

Al-Hazan (965–1040)

Al-Hazan was a pioneer of optical science. The Greeks thought that vision was possible because the eye emitted light. Al-Hazan argued against this, saying that vision occurred because of light rays reaching the eye. His work on optics was translated into Latin and was very influential on European scientists, including Kepler.

Al-Hazan had a theory as to why the moon appears larger when it is near the horizon. He thought that the phenomenon was an optical illusion and that it is objects on the horizon that trick us into thinking the moon is bigger.

Al-Biruni (973–1048)

Al-Biruni was a polymath and world-class scientist. His name is not well-known in the West but he can be considered an Islamic da Vinci. Al-Biruni knew that the sun was at the centre of the cosmos and that the earth rotated around its axis. (It would be 600 years before Copernicus had the same thought.) He wrote books on many subjects, including medicine, geography, physics, astronomy, theology and astrology.

Nasir al-din al-Tusi (1201–74)

Nasir al-din al-Tusi was another Muslim astronomer to point out some serious shortcomings in Ptolemy's work. He wrote *A Memoir on Astronomy*, in which he devised a new model of lunar motion completely different from Ptolemy's. Al-Tusi's other work, *Zij-i ilkhani* (*Ilkhanic Tables*), was a bestseller among astronomers until the fifteenth century.

Ulugh Beg (1393–1449)

Ulugh Beg was the grandson of Tamberlaine. He lived and ruled in Samarkand, which he turned into a great intellectual centre. He was a famous astronomer and built a huge observatory, which sadly was destroyed soon after his death.

Astronomers, fifth century BC–AD 1450

Era	Date	Astronomer	Achievements
Babylonian	Fifth century BC	Nabu-Rimanni	Devised tables for calculating position of the sun, moon and stars
Babylonian	Fourth century BC	Kidynas	Measured length of the lunar month
Babylonian	Third century BC	Sudines	Passed Babylonian astronomical knowledge on to the Greeks
Greek	c.610 BC	Anaximander	Saw that the moon reflected light from the sun
Greek	625–546 BC	Thales	Thought the earth flat
Greek	c.580 BC	Pythagoras	Believed the moon was a planet, which he called 'counter-earth'
Greek	500–428 BC	Anaxagoras	Had a theory about eclipses and relationship of the earth, sun and moon

»

Era	Date	Astronomer	Achievements
Greek	c.423 BC	Meton	Calculated that the moon returned to the same spot once every 19 years (Metonic cycle)
Greek	408–347 BC	Eudoxus	Thought the sun, moon and planets were held in 27 spheres with the earth at the centre
Greek	c.370 BC	Callipus	Calculated the Callipic cycle at 76 years – I day. More accurate than the Metonic cycle
Greek	310–230 BC	Aristarchus	Thought the sun at the centre of the universe
Greek	Mid second century BC	Hipparchus	Thought to be a genius. Put the earth at the centre of the universe. Influenced Ptolemy
Greek	AD 85–165	Ptolemy	Put the sun at the centre of the universe. Ideas dominated for nearly 2000 years
Islamic	858–929	Albategnius	Wrote the very influential *On the Motion of the Stars*, influenced Galileo and Copernicus
Islamic	965–1040	Al-Hazan	Had a theory as to why the moon seems bigger on the horizon
Islamic	973–1048	Al-Biruni	A very great scientist and polymath. Put the sun at the centre of the universe
Islamic	1201–74	Nasir al-din al-Tusi	Had problems with Ptolemy's geocentric universe
Islamic	1393–1449	Ulugh Beg	Ruler and astronomer. Built finest observatory of his time in Samarkand

Astronomy in the West
AD 1100 to Present Day

The influence of Islamic thought on astronomy declined and by about 1100 scientific thought began to re-emerge in Western Europe. The known world was expanding and the need for an accurate navigational system and especially a way of determining longitudinal position became acute. Ships travelled further and further afield. The church demanded an accurate chronology to compute religious festivals such as Easter. Astronomy became important. The Islamic translations of Ptolemy and other great classical thinkers began to filter into Europe. A great revolution was in the offing.

In 1543 Copernicus published his treatise *De Revolutionibus Orbium Celestium* in which he propounded his theory that the sun and not the earth was at the centre of the universe. This view overturned more than 1500 years of conventional wisdom. At the same time the idea that knowledge could be gained through repeated experiment, observation and deduction, what we call science, slowly took hold.

Tycho Brahe (1546–1601) built a lavish observatory on the Danish island of Hven, where he designed the largest and the most accurate astronomical instruments yet seen. Tables of logarithms appeared at the beginning of the seventeenth century and greatly simplified the business of multiplication and division, essential requirements for the astronomer. The most important development of all was the arrival of the telescope.

Over the next 400 years the telescope developed into a powerful tool. By the nineteenth century two more major technological advances had been made: the discovery of electricity and the development of photography. Photography made it much easier to record what the telescope revealed. Electricity made it possible to see things hidden even to the telescope.

The telescope encouraged people to think about the nature of light. Newton discovered the spectrum and later it was discovered that light travels in waves vibrating at different frequencies. Even later came the exciting discovery that some light waves are invisible to the human eye. Electricity made it possible to look at the invisible frequencies and to speculate about the composition of the object from which they came. This is called spectroscopy, the study of electromagnetic radiation. We can now study light across a very broad spectrum including infra-red, ultra-violet, X-ray, nuclear magnetic and many more. Most major observatories have spectroscopes working alongside telescopes.

The work on optics was often carried out by men who were using telescopes to study the moon. These include Francesco Maria Grimaldi (1618–63), who made observations about the defraction of light; Christiaan Huygens (1629–95), who discovered that light moved in waves; and the much neglected Robert Hooke (1635–1703), who worked on defraction and wave theory.

Astronomy became more complex as the boundary between physics and astronomy blurred. Johannes Kepler discovered the three laws now known as Kepler's Laws of Planetary Motion. Galileo is well-known for his assertion that the universe is heliocentric. It is less well-known that he described a theory of relativity that was the basic framework of Newton's laws of motion and which is central to Einstein's special theory of relativity.

The foundations of modern physics were laid when Isaac Newton realized that it was gravity that kept the moon in orbit and went on to describe the three laws of motion.

Apollo 11 made its precarious way to the Sea of Tranquillity using theories of gravity and motion that had flickered into life hundreds of years before and which had been developed by men who never rose further from the earth than they could jump.

SOME IMPORTANT FIGURES IN THE HISTORY OF WESTERN ASTRONOMY

Nicolaus Copernicus (1473–1543)

Copernicus was a diplomat, physician, military leader, philosopher, astronomer and mathematician. His great work *De Revolutionibus Orbium Celestium* (*On the Revolutions of the Celestial Spheres*) describes a universe that has the sun at the centre. After its publication traditional geocentric astronomy was never the same. Copernicus died of a stroke as his book was published. Legend has it that as he lay dying his friends placed the book in his hands. He died peacefully, quietly staring at his life's work.

Nicolaus Copernicus.

Tycho Brahe (1546–1601)

Tycho was the last major astronomer to believe that the earth is at the centre of the solar system. He was rich and used his money to make the finest private observatory in Europe. He set out to observe and record the movements of the heavens. He was helped in this by his sister Sophia. Their measurements were made with great rigour and achieved unprecedented accuracy. Tycho's most eccentric belief was that the earth is static and does not revolve. He lost the bridge of his nose in a duel and for everyday use replaced it with a nose made of copper. On special occasions he wore a nose made of gold and silver.

Galileo Galilei (1564–1642)

Galileo was an astronomer, a physicist and a mathematician. Like Tycho he worked with rigour and the results of his experiments can be analysed with great precision. He pioneered the early development of the refractor telescope and used it to discover the moons of Jupiter. He was one of the first people to study the surface of the moon. He discovered that contrary to popular religious opinion the moon was not a smooth celestial orb but roughened by craters and mountains. His best-known work is *Siderius Nuncius* (*The Messenger of the Stars*) in which he champions Copernicus' assertion that the sun is at the centre of the universe. He did not believe the moon had anything to do with the tides.

Galileo Galilei.

Johannes Kepler (1571–1630)

Kepler was an astrologer, astronomer and mathematician. He worked for a time in Tycho Brahe's observatory. Kepler's most important work was his *Epitome of Copernican Astronomy*, in which he described his three laws of planetary motion. In 1615 his mother was tried and imprisoned for witchcraft. She was released after fifteen months.

Johannes Kepler.

Christiaan Huygens (1629–95)

Huygens was a key figure in the scientific revolution that was taking place all over Europe. His most important contribution was the assertion that light travelled in waves. He discovered that the rings of Saturn were made of rocks. He formulated as a quadratic equation the ideas that became Newton's second law of motion. Newton turned the equation into a general statement. Huygens believed there was life on other planets very similar to that found on earth.

Robert Hooke (1635–1703)

Hooke's reputation has been obscured by Isaac Newton. He was one of the most important scientists of his day. Newton knew of and used Hooke's work on gravity and its relationship to the motion of the planets. Hooke designed several Gregorian telescopes, using them to study and draw the craters on the surface of the moon.

Isaac Newton (1643–1727)

Newton was a polymath with interests in almost everything that could entice the enquiring mind. He was especially obsessed with mathematics, alchemy, theology and astronomy. His book *Philosophiae Naturalis Principia Mathematica* (*Mathematical Principles of Natural Philosophy*) has been described as the most influential work in the history of science. In 2005 members of the Royal Society rated him above Einstein in importance. He formulated the three laws of motion and designed the first reflector telescope.

Edwin Hubble (1889–1953)

Hubble was an American astronomer who worked mainly at the Mount Wilson Observatory using the 100-inch Hooker telescope. He revolutionized astronomy by indicating that there were other galaxies in the universe as well as the Milky Way. Investigating the phenomenon known as 'red shift', he formulated Hubble's Law, which shows that the speed of a galaxy which is moving away from the Milky Way gets faster the further away it is. This discovery was a cornerstone of the Big Bang theory about the beginning of the universe. Hubble has a crater on the moon named after him. The Hubble Space Telescope also carries his name.

Important astronomers 1400–present day

Culture	Dates	Name	Importance
European	1473–1543	Nicolaus Copernicus	Argued that the sun was at the centre of the universe
European	1546–1601	Tycho Brahe	Last of the great pre-telescope astronomers. Put the earth at the centre of the universe
European	1564–1642	Galileo Galilei	Second man to look at the moon through a telescope. Supported Copernicus, put the sun at the centre of the universe
European	1571–1630	Johannes Kepler	Devised three laws of planetary motion, supported Copernicus
European	1629–95	Christiaan Huygens	Believed in life on other planets. Asserted that light was made up of waves
European	1635–1703	Robert Hooke	Very important. Overshadowed by Newton, who used his ideas
European	1643–1727	Isaac Newton	Invented a reflecting telescope. Described three laws of motion. Voted most important scientist ever
American	1889–1953	Edwin Hubble	Proved that there were other galaxies as well as Milky Way. Contributed to proof of what is now the Big Bang theory of the creation of the universe

The Telescope and the Moon
Galileo to the
Twenty-First Century

Lenses have existed for a long time. Examples have been found in Crete and Asia Minor which are thought to be 2000 years old. A lens dating from 650 BC was found in an archaeological dig at Nineveh. The Greeks and the Romans knew about and probably used lenses. The word itself comes from the Latin for lentil.

The first recorded telescope patent was applied for by Hans Lippershey (1570–1619) in the Netherlands in 1608. Legend has it that the telescope was invented by accident when one of Lippershey's apprentices held two lenses in line with each other and noticed they caused things to appear bigger. Lippershey mounted the two lenses in a tube which he termed an 'optic tube'.

Legend has it that the telescope was invented by accident when one of Lippershey's apprentices held two lenses in line with each other and noticed they caused things to appear bigger.

The astronomer Galileo heard about the optic tube, made one of his own, pointed it at the moon and made his famous drawings of the moon's surface.

Galileo's telescope had two lenses 1.6 inches in diameter giving a 3× magnification. Today one of the largest telescopes in the world has optical elements nearly 33 feet in diameter. It is mounted at the W. M. Keck Observatory on Mauna Kea in Hawaii. It weighs 300 tons, can be moved with microscopic precision and has examined objects 40 million light years away. Other optical telescopes are being planned that have elements 300 feet in diameter.

TELESCOPE DESIGN

There are two sorts of telescope: refractors and reflectors. The light-gathering element of a telescope is known as the objective. Refractors use glass lenses for the objective and reflectors use mirrors. Galileo's telescope was a refractor.

The simple refractor telescope

Refractors are very good telescopes. They have two serious drawbacks: they cannot support large lenses and they suffer from chromatic aberration.

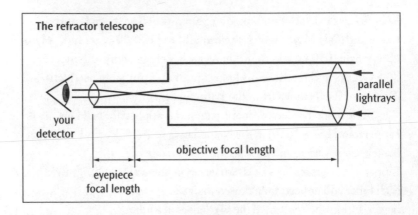

The refractor telescope

parallel lightrays

your detector

objective focal length

eyepiece focal length

Supporting the lens

A lens can only be supported round its edges. If the lens gets very large the weight of the glass causes the lens to sag in the middle. Today the largest refractor telescope has a lens with a diameter of 40 inches.

Chromatic aberration

White light is made up of a spectrum of different-coloured lights. The light entering the lens at the front of the telescope is bent or refracted by the glass, which causes the white light to break up into its colour components. This causes distortion of the image, the symptoms of which are fringes of colour round the edges.

The Types of reflector telescope

Reflector telescopes gather and focus light with a curved mirror. The mirror can be supported across its back and is not prone to sagging. The light is not refracted and the telescopes do not suffer from chromatic aberration. Reflector telescopes can be very large. The telescope at the W. M. Keck Observatory has a mirror with a diameter of 33 feet. It would be very difficult and unnecessary to make an equivalent optically perfect lens. The mirrors in reflectors do just as good a job as a lens and are easier to make.

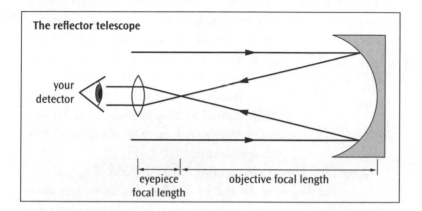

Spherical Aberration

Refractor and reflector telescopes both suffer from spherical aberration, which is a phenomenon that occurs when the lens or the mirror is not curved correctly. This prevents the exact focus of light entering the telescope. Both spherical and chromatic aberration can be eliminated by using corrective lenses. Corrective lenses cut down the light that passes through the telescope to the observer, which corrupts the image.

DEVELOPMENT OF THE TELESCOPE

Refractor telescopes were made more powerful by increasing the focal length of the lenses. As a result telescopes became longer and more unwieldy. Johannes Hevelius (the author of the *Selenographia*) had a telescope that was 150 feet in length. It was suspended from a 90-foot pole and needed a team of ten men to raise and lower it.

In 1668, Sir Isaac Newton built the first reflector telescope. It was 6 inches long, had a 40x magnification and was able to do the work of a refractor six and a half feet long.

The history of the telescope became a competition between the refractor and the reflector. As time passed, glass technology improved, mirror-making improved and the understanding of optics improved. Until the end of the nineteenth century, the finest telescopes were refractors. Eventually the refractor reached a size beyond which it could not go. On 9 June 1897, at the Yerkes Observatory, the largest refractor telescope ever built went into operation. The lens is 40 inches in diameter and weighs 500 pounds. The telescope tube is 60 feet long and the whole instrument weighs 20 tons. It is still operational.

Until the end of the nineteenth century, the finest telescopes were refractors.

After 1897 it was reflectors that ruled the world of optical astronomy, but they were dwarfed by the emerging technology made possible by the development of electricity and the discovery of radio waves. Radio astronomy made it possible to build huge non-optical telescopes. At the Arecibo Observatory in Puerto Rico there is a radio telescope built into a volcano. It is 1000 feet wide.

Important events in the development of the optical telescope in Europe and the USA

❭ **1608, Denmark**
Hans Lippershey applies for first telescope patent
In 1608, spectacle maker Hans Lippershey applied for a patent on a telescope and produced the first commercial model. It was known as the Danish perspective glass and had a magnifying power of 3x.

❭ **1610, Italy**
Publication of first telescope-aided drawing of the moon
In 1610, Galileo observed the moon through a refractor telescope and published his observations in *Siderius Nuncius* (*The Messenger of the Stars*). Galileo was the second man to make such a map. The first man to draw the moon through a telescope was Thomas Harriot in England. His drawings were made three months before Galileo's but published later than *Siderius Nuncius*.

1611, Bavaria
Johannes Kepler publishes Dioptrice

Kepler's *Dioptrice* was a work on optics in which he described the causes of the phenomenon of spherical aberration and suggested a solution. Kepler also established that light proceeds from an object towards the eye. Up to then it had been thought that the eye emitted light. Kepler designed but did not make a telescope using the same lenses as Galileo but repositioned the eyepiece further from the objective. This made a better telescope, but the image was inverted. All telescopes since then have had inverted images.

Through the seventeenth century technological advances allowed for more powerful refracting telescopes. At the same time they become longer and more unwieldy. This continued up to the beginning of the eighteenth century, when long refracting telescopes reached their zenith. The seventeenth century is the age of the long telescope. In 1733 the achromatic lens was invented, which made it possible to build shorter refractor telescopes.

1612, Italy
Word 'telescope' coined

A Greek mathematician, Giovanni Demisiani, coined the term 'telescope'.

1638, England
William Gascoigne introduces micrometers into telescope design

A micrometer is an aid to measuring the movement of, and the distance between, stars. They made telescopes easier to use when making astral measurements.

1663, England
Isaac Newton discovers the spectrum

Newton discovered that white light is made up of a spectrum of coloured light. Lenses bend white light, causing it to break up into its component colours. This is called chromatic aberration. The mathematician James Gregory (1638–75) attempted to design and build a reflecting telescope that eliminated chromatic and spherical aberration. He failed because mirror and glass technology was not developed enough. But his ideas were sound, and Gregorian reflector telescopes built since his time have worked perfectly.

1668, England
Isaac Newton builds the reflector, a revolutionary new telescope
Newton designed and built a reflecting telescope with a magnification of 40x. The instrument was 6 inches long and could do the work of a six-and-a-half-foot long refractor.

1674, England
Flint glass
A superior type of highly transparent glass called flint glass was developed by leading glass maker George Ravenscroft (1618–81).

1721, England
George Hadley's Gregorian reflector
Hadley presented the Royal Society with a Gregorian reflector he had built. The telescope had a reflector 6 inches in diameter and was 6 feet long. The equivalent refractor would be 123 feet long.

1733, England
Chester Moore Hall (1703–71) uses flint glass
Hall designed an achromatic lens using the new flint glass. The lens reduced chromatic aberration. It was made from two different types of glass: a concave lens made out of flint glass and a convex lens made from crown glass. The lens brought light to a focus more quickly and meant that the refractor telescopes could be shorter.

From 1757, the long refractor died out. All refracting telescopes were made with achromatic lenses and are still so made today.

1772, England
William Herschel (1738–1822)
Herschel started construction of large telescopes and modified Newton's telescope design. He also wrote a paper about the mountains of the moon.

1789, England
Herschel's 48-inch reflector
The telescope was huge and very cumbersome.

1824, Russia
Joseph von Fraunhofer's 9.5-inch refractor at Dorpat
The telescope was 14 feet long and
beautifully made and mounted.
It was easily manoeuvred and was known
as the Great Dorpat Refractor. It was the
biggest refractor telescope in the world.

It was the biggest refractor telescope in the world.

1842, England
William Parsons, 3rd Earl of Rosse (1800–67), orders construction of a 72-inch reflector
This was the biggest reflector telescope in the world. The mirror weighed 4 tons.

1888, USA
36-inch refractor erected at the Lick Observatory

1897, USA
40-inch refractor commissioned at the Yerkes Observatory
This is still the largest refractor in the world.

1993, USA
400-inch reflector erected at W. M. Keck Observatory
The telescope, whose lens is 33 feet across, is part of a complex of instruments that can observe the universe across a spectrum of wavelengths that include visible light, near infra-red and infra-red.

There are many types of telescope in operation today. They can scan many wavelengths. They can be mounted on probes and sent on journeys that will take many years. If telescopes are mounted on the moon they will take advantage of the satellite's unpolluted atmosphere and low gravity to look deeper and deeper into space and time.

Important events in telescope design		
1608	Hans Lippershey applies for telescope patent	Denmark
1610	Galileo's drawing of moon as seen through 1.6-inch refractor telescope	Italy
1638	William Gascoigne introduces micrometer into design to enhance accuracy	England
1663	Isaac Newton discovers spectrum	England
1668	Isaac Newton designs reflecting telescope	England
1674	Superior flint glass developed	England
1789	William Herschel builds 48-inch reflector	England
1842	William Parsons, 3rd Earl of Rosse, orders 72-inch reflector	England
1888	36-inch refractor built at Lick Observatory	USA
1897	40-inch refractor built at Yerkes Observatory Largest in the world	USA
1993	400-inch reflector built at W. M. Keck Observatory	USA

The Camera and the Moon

In 1844 *The Pencil of Nature*, by William Henry Fox Talbot, appeared. It was the first book ever to be illustrated with photographs. The first photographic image of the moon was taken in 1840 by an American, John Draper. Draper used a 12-inch telescope and a twenty-minute exposure.

Camera technology improved and by the end of the century M. Loewy and P. Puiseux were in the process of publishing their *Atlas Photographique de la Lune*. Through the twentieth century camera equipment got lighter and more sophisticated; film became more sensitive. By the end of the century, digital photography started to take over from emulsion-based imagery.

In the fifties, space exploration made its own demands. Special photographic equipment was developed for inclusion in the payload of lunar probes. This included cameras that could robotically

process and transmit images from celluloid. Video cameras were used on Apollo 7 to transmit live pictures from space. Hand-held cameras were carried by astronauts on to the moon's surface.

One of the first photographs of the moon, taken in the early 1880s by Henry Draper.

There are now tens of thousands of close-up images of the lunar surface.

The twentieth century closed with the publication of *The Clementine Atlas of the Moon*, a work compiled from images taken by the probe Clementine.

Mapping the Moon

Map making is a difficult science. Serious map making on earth only really became possible in the eighteenth century, with the development of optical measuring equipment. Even then there were problems. The earth is very big and many parts of it are hard to get at or dangerous. Before space travel no earth-based cartographer could step back and see the big picture.

No such worries hampered the lunar map-maker. The moon is there, plain for all to see, and it is not dangerous to look at. The big problem is that you can only see half of it, you cannot see it all the time, and it is a very long way away. Lunar map making took off with the invention of the telescope. Each new feature that was revealed had to be positioned on the map and catalogued. The problems of naming grew. And there were always the vagaries of the weather to contend with. The finest optical telescope in the world cannot see through clouds and is susceptible to heat and cold.

HISTORY OF MOON MAPPING

At the beginning of the seventeenth century, William Gilbert drew the first map of the moon. He used his naked eye to identify and name thirteen lunar features. By the end of the nineteenth century, a map had been published which attempted to identify and name nearly 33,000 features.

More than 250 years separated the two documents. In that time three or four major systems of naming had been used, as well as countless minor ones. With time lunar feature naming became a sort of scientific trainspotting. Many maps acquired a fantastic and unreadable complexity.

The first major map of the moon was published in 1645 by Michiel van Langren. Van Langren used the names of royalty, nobility, explorers and scientists to identify features. Like Gilbert, he used Latin names for land masses and supposed water.

In 1647 another major map appeared, drawn by Johannes Hevelius. Hevelius used a cumbersome system of Latinized land-mass names based on earthly counterparts. Of his many proposed feature names, which include 'chersonnesus', 'celenorum' and 'corocondametis', only ten have survived on to modern maps.

In 1651 two Jesuit priests, Giovanni Riccioli and Francesco Grimaldi, published a map that combined some of van Langren's names with the names of important figures who were connected with astronomy. Like others before them, the two priests used the generic names of topographical features found on earth.

The two maps by Hevelius and Riccioli/Grimaldi became the standard works for the next 150 years. During that time new names were added and lost; Greek and Roman letters were added; and numbers were brought in to help. In the mid nineteenth century the camera further complicated the picture. Lunar map making and chaos became the same thing.

In 1919 the International Astronomical Union (IAU) was set up to try to regularize the process. Committees and problems proliferated throughout the twentieth century. It was difficult to agree even basic questions of map orientation. The telescope presents the observer with an image which is upside down, 'south up', and maps had been drawn to reflect that. Astronauts wanted a more conventional 'north up' map. The question had to be solved by a vote of the assembly of the IAU, which decreed that moon maps should be printed with north at the top. Some astronomers continue to use the telescopic convention and put south at the top.

In 1960 a whole new set of problems arrived. The Soviet probe Luna 3 flew round the moon, photographing the far side. This was an epoch-making event, which the Russians were quick to take advantage of. They immediately published *Atlas of the Far Side of the Moon*. It contained names such as Mare Moscovrae.

The major powers, the United States, Europe and the Soviet Union, struggled to find a solution. In 1961, the IAU set up Commission 16a for Lunar Nomenclature and Cartography. At the far end of a very dark tunnel there was a glimmer of light. After much debate and a lot of bad temper, systems for plotting lunar features were standardized. Finally, in 1982 NASA published the *Catalog of Lunar Nomenclature*. This was a definitive list of all the internationally agreed features. It was complete to 1981, and it is still being updated today.

IMPORTANT LANDMARKS IN LUNAR MAPPING FROM 1600 UNTIL THE INVENTION OF PHOTOGRAPHY

1600
William Gilbert's moon map
About 1600, William Gilbert (1544–1603) drew the first known 'naked eye' map of the moon for inclusion in his book *De Mundo Nostro Sublunaria Philosophia Nuova* (*A New Philosophy of our Sublunar World*). Gilbert imagined the moon to be like the earth, with vast land masses surrounded by oceans. He gave the moon earth-style names using medieval scientific Latin. Thanks to Gilbert, the moon's surface sprouted *Mare*, *Isole*, *insula* and

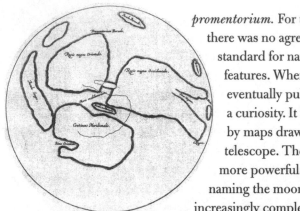

promentorium. For the next 400 years, there was no agreed scheme and standard for naming the moon's features. When Gilbert's map was eventually published in 1651 it was a curiosity. It had been overtaken by maps drawn with the aid of a telescope. The telescopes became more powerful, and the problem of naming the moon's features became increasingly complex and confused.

1610

Galileo's *Siderius Nuncius* (*Messenger of the Stars*)

Galileo did not attempt to name the moon's features, but he did state that it was an earth-like body with areas of earth-like land masses. He did not claim that there were seas on the moon, but he did say that the oceans of the earth would appear as dark areas to an astronomer, and that the moon had similar dark areas. These ideas caught on. Today the rocky dry moon has features named as seas and oceans: Mare Serenitatis, Oceanus Procellarum, etc.

1611

Thomas Harriot

Thomas Harriot drew the first optically aided pictures of the moon on 5 August 1609, several months before Galileo drew his. The Galileo pictures were published one year before Harriot's and it was Galileo who was credited with being the first man to point a telescope at the moon's surface. Harriot's image was 6 inches in diameter. He attempted to identify the moon's features with a system of letters and numbers.

Between them, Galileo and Harriot observed the phenomenon of the moon's librations. Librations are caused by the fact that the moon is slightly irregular in its orbit. From an observer's point of view it wobbles as though the moon is a head which is nodding and turning from side to side. The result is that we can see 59 per cent of its surface, though not all at the same time. Different bits round the edge of the moon's sphere are visible at different times. Drawing this on a map proved to be a continual problem.

1637

The Mellan Engravings

The engraver Claude Mellan worked under the instruction of two astronomers, Pierre Gassendi and Nicolas de Peiresc. They used a telescope, the optics of which were provided by Galileo. Mellan drew three phases of the moon: waxing gibbous, waning quarter and full. His work was published in early 1637 and is of outstanding beauty. Gassendi and Peiresc had planned this as just part of a larger work which would, among other things, help with the problem of establishing longitude for navigators on earth. Peiresc died in the year the map was published and the project foundered. Gassendi had attempted to name the features on the moon, but only his notes remain and they are very difficult to link to the map as there are no guide marks as to where they apply. He had planned to use a system of letters and numbers, combined with the Latin names of earth's topographical features such as *Mare*, *Vallis*, *Rupes* and *Mons* (sea, valley, cliff and mountain).

The first moon map, engraved in 1635.

1645

Michiel van Langren's lunar map

Michiel van Langren (1600–75) was a member of an important Flemish globe- and map-making family. In the early 1630s, van Langren set out on a major lunar project. He had the financial support of King Felipe IV of Spain's aunt, Princess Isabella. He planned to publish a theory of the moon's librations, drawings of lunar phases and a map. The map would be a moon globe with latitude and longitude circles marked, along with instructions for obtaining longitudes from sunset readings on listed lunar features. This was a very complicated plan. By about 1645 he had completed thirty drawings and established a naming system which used the names of royalty, illustrious men and saints. His image was 34 centimetres in diameter and was published in March 1645. There are only four known copies of this map in the world and they have three slightly different forms. The earliest is thought to be the van Langren map held in the Leiden Observatory. The next, slightly different from the first, is in the Royal Observatory in Edinburgh. The last two, which are the same, are in the Bibliothèque Nationale, Paris, and the Observatorio de Marina in San Fernando. There is a further copy – a forgery – in the library of the University of Strasbourg. Many of van Langren's lunar names were linked to the Spanish royal family. They did not catch on.

> *There are only four known copies of this map in the world and they have three slightly different forms.*

1647

Johannes Hevelius' *Selenographia*

Johannes Hevelius (1611–87) came from an affluent brewing family. From 1643 he made observations and drawings of the moon. His work was published in 1647 as *Selenographia*. This large volume became one of the two major sources of lunar information for the next 150 years (the other being the map by the priests Riccioli and Grimaldi). The *Selenographia* contained a lot of technical information about the moon, the sun and Jupiter. The book's most impressive features are three maps of the full moon and forty drawings of the moon in its various phases. In the book, Hevelius asserts that the bright areas of the moon are land and that the dark areas are water.

The three full moon maps were labelled P, Q and R, and each was about 29 centimetres in diameter.

Moon map by Johannes Hevelius, 1654, showing librations.

◗ P showed the full moon with geographical features that would be seen at that phase.
◗ Q had some of the names engraved on to it and had an accompanying table of other names.
◗ R had added features which are visible at other phases of the moon.

Each map had dotted lines to show the areas revealed by the moon's librations.

1651

Giovanni Riccioli and Francisco Maria Grimaldi:
Almagestum Novum

Grimaldi's map borrowed from van Langren, Hevelius and others, and included his own corrections 'made with the best telescope from many phases'. Many of the names are still in use today.

Riccioli (1598–1671) and his friend Grimaldi (1618–63) were Jesuit priests and therefore bound to adhere to the church's dogma that the earth was at the centre of the universe.

Riccioli spent three years attempting to write the most authoritative book on astronomy ever written. This was published

Giovanni Riccioli's moon map of 1651 is the forerunner of the modern system of naming the moon's features.

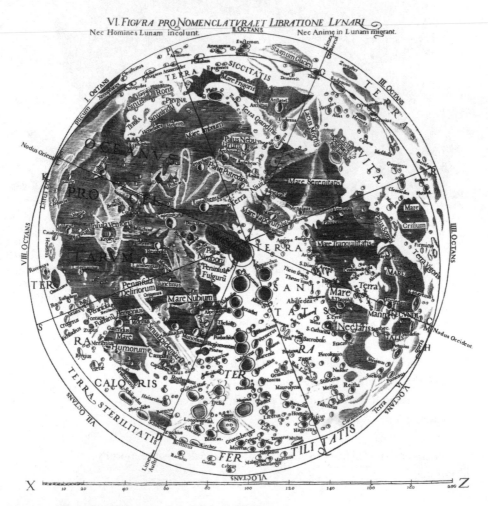

in 1651 under the title *Almagestum Novum*, the *New Almagest*. It was a successor to the *Almagest* of Ptolemy, whose theories about the earth-centred universe had dominated scientific and spiritual thought for 1500 years.

The book contained a section on the moon which was illustrated with two maps drawn by Grimaldi. Riccioli introduced a completely new method of naming the moon's features. To name the seas, continents, lakes and peninsulas he used descriptions of natural and psychological states: e.g., Palus Putredinis (Marsh of Decay), Sinus Iridium (Bay of Rainbows) and Peninsula Deliriorum (Peninsula of Insanities).

To name the craters he divided the moon into eight regions and used the names of astronomers, philosophers and saints – e.g., Keplerus, Plato and St Catharina. We therefore have Copernicus in the Sea of Storms (Oceanus Procellarum) and Aristotle in the Sea of Cold (Mare Frigoris). Riccioli's system has remained, with additions, almost intact to the present day.

The first director of the Paris Observatory, Jean Dominique Cassini (1625–1712), used Riccioli's name system for one of his own maps, and quite soon after that Hevelius' maps were being published with Riccioli's names attached to them. Three hundred years later, Apollo 11 landed in Riccioli's Mare Tranquilitatis (Sea of Tranquillity).

It is possible to analyse the way Riccioli laid out his names on the moon's surface and find evidence that he was not in his heart convinced that the earth lay at the centre of the universe. Great prominence and physical proximity is given to Aristarchus, Kepler and Copernicus, the three most famous astronomers who believed that the sun was at the centre of the universe. It was the work of Copernicus that had finally overturned the work of Ptolemy, ending a 1500-year belief that the universe revolves around a static earth.

The maps of Hevelius and Riccioli/Grimaldi dominated lunar cartography for the next 150 years.

1775

Tobias Mayer's moon map
Tobias Mayer (1723–62) had been dead for thirteen years when his map was published. He produced two identical maps with one at twice

the scale of the other. They were published in 1775. Mayer was the first astronomer to realize that the moon's features could be accurately plotted using trigonometry. The publisher of the map, George Lichtenberg, had the novel idea of adding longitudinal and latitudinal grid lines to the map, something which had never been done before. The map was more accurate than any of its predecessors.

1791

Johann Hieronymus Schröter:
Selenotopographische Fragmente

Johann Hieronymus Schröter (1745–1816) was an amateur astronomer. His ambition was to observe, draw and measure lunar features under all conditions of light. The observatory that he worked in had five telescopes: a 2.5-inch refracting telescope by Dolland; a 4.75-inch and 6-inch reflector by W. Herschel; and a 9.5-inch reflector by Schrader. He also constructed an 18.5-inch reflector himself, which was a very clumsy and inefficient instrument. Schröter devised too what he called a *Projections-Maschine* with which he could look at the moon with one eye and at his drawings with the other. The results were impressive and were published in two parts in 1791 and 1802. The two volumes had more than seventy-five plates, the equivalent of a lunar map 4 feet across. The publication included a new engraving of Mayer's map, added seventy-six new features and introduced the word crater into lunar map making. Schröter used Riccioli's names on the drawings but used both Riccioli's and Hevelius' names in the text. He also used a system of letters to identify other lunar features.

The publication included a new engraving of Mayer's map, added seventy-six new features and introduced the word crater into lunar map making.

His atlas has been described as the best made before the introduction of photography. Sadly for Schröter, the French occupied Lilienthal in 1813 and his observatory, complete with telescopes and notes, was ransacked and burnt to the ground. His life's work was reduced to ashes. Schröter died three years later at the age of seventy-one.

1805
John Russell (1745–1806)

Russell published two very detailed lunar maps, which were the fruit of forty years' work. He began his project in 1764. The maps were about 14 inches in diameter and were the most detailed yet produced. Like Schröter's work, Russell's maps would not be surpassed in clarity and detail until photography became available as a map-making tool. In 1797, Russell produced a globe mounted in an elaborate brass mechanism which he called *Selenographias*. Very few copies of the globes or the maps were produced. One can be seen at the National Maritime Museum in Greenwich and another is in the Museum of the History of Science in Oxford.

1837
Beer and Mädler, *Mappa Selenographica* and *Der Mond*

Between 1828 and 1837, Johann Heinrich von Mädler (1794–1874), with his friend Wilhelm Beer (1797–1850), set about writing the definitive work on the moon. Equipped with a 3.75-inch refracting telescope which belonged to Beer, they planned a map about 37 inches in diameter. This was to be accompanied by a text detailing everything that was known about the moon. The map was published in four quadrants in 1834 and 1836 under the title *Mappa Selenographica*. This was followed in 1837 by the accompanying descriptive work called *Der Mond*. The book detailed almost every known fact about the moon, and speculated about its origins and structure.

Mappa Selenographica: *southeast quadrant of the Beer and Mädler moon map, 1834.*

The 400 pages of small print had an enormous impact on lunar studies. The two men made the spectacular assertion that the moon was a lifeless dead body. Lunar astronomers felt there was nothing left to discover, that the future would consist only of refining and perfecting the Mädler map. In 1837 Mädler produced a smaller version of his map. This was a general chart, about one third the scale of the original, with 368 features.

OTHER INTERESTING MOON MAPS AND EVENTS, 1600–1900

1661

Sir Christopher Wren's globe

Christopher Wren (1632–1723) produced the first globe of the moon. It was constructed by Joseph Moxon and placed in the King's cabinet of curiosities. Sadly, it was sold and is now lost.

1664

Robert Hooke's crater speculations and moon details

Robert Hooke (1635–1703) was a scientific pioneer. Hooke wanted to work out how the moon had been formed. He speculated on the formation of craters and performed experiments with bullets and clay to arrive at an impact theory of crater formation. He performed similar experiments with water vapour and heated alabaster to arrive at a theory of volcanic crater creation. He also made detailed drawings of the crater Hipparchus.

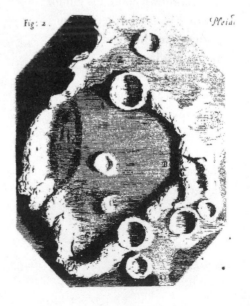

1671

Fr Chérubin d'Orléans (1613–97)

In 1671, Fr Chérubin published a work on optics which contained plagiarized versions of the Hevelius map. He claimed to have devised and used a rhombic pantograph to make enlarged drawings of lunar details – unlikely given the technology of the time.

1679

J. D. Cassini (1625–1712)

With the help of two assistants at the Paris Observatory, Cassini made a detailed lunar map roughly 54 centimetres in diameter. He also produced drawings of lunar features. Over fifty of the plates he produced are at the Paris Observatory. Prints from the copperplate are very rare.

The Maiden in the Moon *from one of Cassini's moon maps, 1679.*

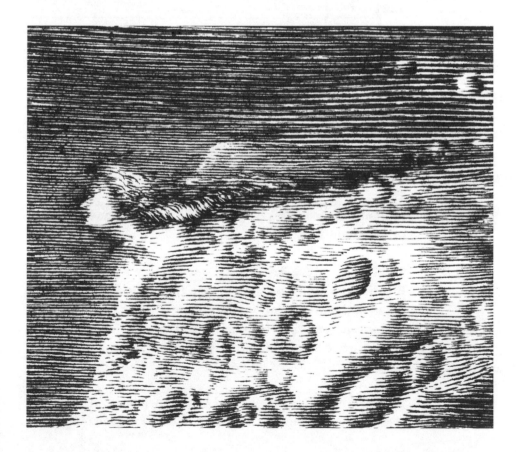

In 1692, Cassini also produced an image of the full moon which he hoped would aid study of the total eclipse predicted for 28 July 1692. This map was copied, reproduced and added to for the next sixty years. It became the standard image of the moon in the French astronomical almanac *Connaissance des temps*. Some of Cassini's details were fanciful, such as the lady's head which he drew as an illustration of the feature *Promontorium Heraclides*.

1685
Discovery of lunar details by Christiaan Huygens (1629–95)
Huygens was another scientific pioneer who made several sketches of unusual lunar details. His work went undiscovered for a century and was not published for 200 years.

1821
Franz von Gruithuisen
Gruithuisen studied the moon and made many studies of small areas. He believed that the moon was inhabited and drew pictures of what he believed to be a lunar city lying between the craters Mösting and Eratosthenes. He was ridiculed for his theory even by astronomers who shared his belief that there was life on the planet. Gruithuisen also produced an incomprehensible scheme for identifying lunar features.

1853
In Bonn, Thomas Dickert produced a 19-foot model of the moon's near side

1865
British Association resolves to map the moon in great detail at 200 inches per moon diameter
The scheme was placed in the hands of William R. Birt (1804–81) and foundered because the methodology was too complex.

1874
James Naysmith (1808–90) and James Carpenter (1840–99)
Naysmith was the inventor of the steam hammer. He and Carpenter

produced a book, *The Moon*, which was devoted to examining the nature and origin of the moon's surface. Later editions had photographs of plaster models of the surface. Naysmith and Carpenter believed that the craters were volcanic in origin.

1895

T. G. Elger (1838–97), *The Moon*

Elger produced a detailed guide to the moon. It was accompanied by a map available as one large sheet or four smaller quadrants. The book has remained in print ever since. It is currently available in paperback on Amazon.

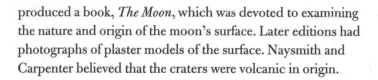

IMPORTANT EVENTS IN LUNAR MAPPING, 1894 TO THE PRESENT DAY

1894–1910

M. Loewy and P. Puiseux: *Atlas Photographique de la Lune*

Between 1894 and 1910, M. Loewy (1833–1907) and his colleague P. Puiseux (1855–1928) undertook to make a photographic survey of the moon. They used a new telescope, the 24-inch Equatorial coudé. They produced about 6000 plates and used the eighty best shots for their *Atlas*. The negatives were blown up to 2 x 3 feet and were enlarged using a process known as *héliogravure*.

1913

Publication of maps by Julius Franz and Samuel Saunder

Working separately, these two men attempted to produce the most detailed maps of the moon yet realized. Franz worked on the areas towards the edge of the moon (the limbs), while Saunder concentrated on the main surface area. There was no international body to regulate the naming of new features, which were being discovered all the time. In 1905, Saunder brought the problem to the attention of the Royal Astronomical Society. In 1910, a committee with international representation was set up

Working separately, these two men attempted to produce the most detailed maps of the moon yet realized.

under the chairmanship of Herbert H. Turner, Director of Oxford University Observatory.

The committee asked itself whether the moon should be named according to a modified traditional naming system or whether a new system should be introduced. Saunder argued for a traditional system. He suggested that a new map be drawn. The map would use the traditional names from the Beer and Mädler Map. Saunder also wanted accepted new names to be retained; capital Roman letters to be used for 'craters, depressions and dark areas'; and small Greek letters for 'peaks and bright spots'. Saunder would map the inner regions and Franz the outer regions, which are subject to the moon's librations.

The committee asked itself whether the moon should be named according to a modified traditional naming system or whether a new system should be introduced.

The work was completed by 1913. Franz drew a separate map of the librations, which he published as *Die Randlandschaften des Mondes* (*The Peripheral Regions of the Moon*). The traditional longitude and latitude lines were dropped in favour of X and Y coordinates, which became known as the xi/eta system.

1913
Collated List of Lunar Formations: Mary Blagg
In order to produce their maps Saunder and Franz had to collate and coordinate the names from the three main nineteenth-century maps, as well as names from other sources. This huge task fell to Mary Blagg. She worked with limited source material to produce an index with 4789 entries, 14,000 cross references between the three main maps and endless references to other material.

1919
International Astronomical Union (IAU) founded
The union was established to coordinate astronomy on a worldwide basis. Thirty-two commissions were set up; Commission 17 had responsibility for the moon. H. Turner was President, and Mary Blagg was among the commissioners. The work of the Commission continued for sixteen years.

1935

Publication of *Named Lunar Formations*

Named Lunar Formations was an internationally approved document standardizing and collating the previous one hundred years of lunar map making. The work was in two volumes, volume 1 being the catalogue and volume 2 the maps. Much of it is the work of Mary Blagg, who was assisted by Karl Müller (1866–1942). The maps are divided into fourteen sections at a scale of 36.5 inches to the lunar diameter. Some of Mary Blagg's maps were incorporated, though she had never intended these to be anything more than a stopgap. *Named Lunar Formations* ended the work of Commission 17, which was then absorbed into Commission 16: Commission for Physical Observations of Planets and Satellites.

1939–45

Second World War: interest in lunar mapping waned

1955

International Astronomical Union proposes photographic map of the moon

The Named Lunar Formations map of 1935 was far from perfect. It was hand-drawn and contained anomalies which had their origin in the Collated List of 1913. A new photographic map was proposed by Commission 16 under the chairmanship of Gerard P. Kuiper (1905–73). Kuiper was director of the Yerkes and Macdonald Observatory. After the Second World War, Kuiper was one of the few astronomers interested in lunar mapping.

> *After the Second World War, Kuiper was one of the few astronomers interested in lunar mapping.*

1958

Kuiper proposes two new projects

Photographic Lunar Atlas and *The System of Lunar Craters* were a development of the 1955 proposal for a photographic map of the moon. E. W. Whitaker was put in charge of the *Atlas* and D. G. W. Arthur took charge of *Lunar Craters*.

1960

The Soviet Luna 3 mission returns with photographs of part of the far side of the moon

The Russians launched the first satellite, Sputnik, in 1959. Later the same year Luna 3 flew round the moon photographing the far side. Luna 3 returned to earth in 1960 after 250 days in space. The success of Luna 3 would start a frenzy of interest in the moon and herald a period of mapping confusion that was as complicated as any of the arguments that had occurred in the previous 300 years.

Luna 3 raised old problems in new forms. Traditionally the moon had been shown as it was seen through a ground-based telescope, inverted, with south at the top – 'south up'. How would this be applied to images taken from space vehicles? There was no accepted rule for the naming of major new features.

By the end of 1960 the Soviets published an *Atlas of the Far Side of the Moon.* The map contained the names of eighteen features never before seen.

1961

International Astronomical Union forms Commission 16a for Lunar Nomenclature and Cartography

By the end of the year the International Astronomical Union had approved two resolutions to establish lunar mapping conventions. Resolution 1 recommended that:

▶ Maps for telescopic observation should be printed south up with the terms east and west deleted.
▶ Maps for exploration should be printed in agreement with terrestrial maps with northern orientation.
▶ The metric system should be used for height and distance.

Resolution 2 dealt with naming the moon's surface. It recommended that:

▶ Known features retain their previous names but with some further recommendations for craters, mountain chains and dark sea-like areas.
▶ Mary Blagg and Karl Müller's *Named Lunar Formations* with

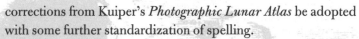

corrections from Kuiper's *Photographic Lunar Atlas* be adopted with some further standardization of spelling.

▶ The names of features on the far side of the moon should accord with the descriptions in the Russian *Atlas of the Far Side of the Moon*.

1961–4
Work starts on *The Lunar Aeronautical Chart* and *The Rectified Lunar Atlas*

The Lunar Aeronautical Chart was a comprehensive programme undertaken by the US Air Force Chart and Information Center to map the moon at a scale of 1:1 million.

The Rectified Lunar Atlas was to be a vertical view of all aspects of the moon's near side (an astronaut's point of view). As these two projects got under way, work continued on *The System of Lunar Craters*.

1967
International Astronomical Union adopts *The System of Lunar Craters*

The System of Lunar Craters was a complete revision of the work that had gone into *Named Lunar Formations* of 1935. A modified version of the catalogue with maps was later published under the title *Lunar Designations and Positions*. There now existed an officially approved system of lunar nomenclature. The maps being produced for the *Lunar Aeronautical Chart* adhered to the system.

1967
Chaos at General Assembly of the International Astronomical Union

The Assembly was flummoxed by the appearance of a revised Soviet *Atlas of the Far Side of the Moon*. The atlas included features photographed by Zond 3 as well as the original Luna 3 material. Nearly half the far-side features had been given Russian names.

At the same assembly a huge photomontage of pictures from the US Lunar Orbiter programme was displayed. The exhibition was laid out on the floor under clear plastic and delegates could walk over it in their stockinged feet.

For the next six years arguments about how to name the moon went from bad to worse. The photographic data flow from the Apollo missions and deep-space exploration threatened to overwhelm everything. Names used on the moon were appearing on other planets. Of 138 new names for features on Mars, 134 were already in use on moon maps. It was chaos.

1973
Formation of the Working Group for Planetary System Nomenclature (WGPSN)
This committee was formed to try to bring order to the situation.

1974
NASA forms Lunar Photography and Cartography Committee
The two new committees began to work together. Names were agreed with the Russians, and E. A. Whitaker devised a 'clock face' system for locating far-side craters. The system treated each large named crater as the centre of a clock. The clock had the letters of the alphabet round its face and smaller craters were named after the letter to which they are closest.

This system has been adopted by the US Geological Survey for its 1:5,000,000 scale maps of the far side. By 1979 the problems of naming had been largely overcome.

1976
The Working Group for Planetary System Nomenclature resolution 1 pt 2
This resolution called for a gazetteer of all the names being used for features on any planet and that this publication should thereafter be updated – a colossal task which would take nearly twenty years to complete.

1982
NASA publishes the *Catalog of Lunar Nomenclature*
The *Catalog* listed the names with map sheet numbers of all lunar features.

1995

Publication of *Gazetteer of Planetary Nomenclature 1994*

The *Gazetteer* listed all the feature names then recognized by the International Astronomical Union. It is continually being updated and can never be complete.

2007

NASA and Google

NASA and Google announced a joint venture to produce new higher-resolution maps and images of the moon.

Key events in lunar mapping 1910–present day		
Date	**Event**	**Details**
1910	*Atlas Photographique de la Lune,* M. Loewy and P. Puiseux	First photographic atlas of moon, 80 2 x 3-foot plates
1913	Maps by Julius Franz and Samuel Saunder. Working to the Committee to Regulate Lunar Naming	Two vols. Longitude and latitude lines dropped in favour of xi/eta system. First attempt to regulate lunar naming
1913	Collated List of Lunar Formations, Mary Blagg	Attempt to coordinate mapping and naming conventions. 4789 entries, 14,000 cross references
1919	International Astronomical Union (IAU) established	Commission 17 of the Union is responsible for the moon
1935	*Named Lunar Formations* issued, an internationally approved publication	Two vols collating work of lunar cartographers over last 100 years
1935	IAU Commission 17 wound up and absorbed into Commission 16 for Physical Observation of Planets and Satellites	
1939–45	Second World War	Second World War destroys interest in lunar naming
1955	IAU proposes photographic map of moon	Gerard Kuiper to chair project

»

Date	Event	Details
1958	Kuiper splits photographic atlas into projects: *Photographic Lunar Atlas* and *System of Lunar Craters*	
1960	Soviet Luna 3 photographs far side of moon	No rules for naming major new features
1961	IAU forms Commission 16a for Lunar Nomenclature and Photography	Proposed: • Telescope maps south up • Exploration maps north up • Suggestions for establishing naming conventions • Recognized Soviet far-side names
1961	USAF starts work on *Lunar Aeronautical Chart* and *The Rectified Lunar Atlas*	Attempt to make an astronaut's POV of the moon
1967	*Atlas of the Far Side of The Moon*, pt 2	Soviet map and huge increase in confusion over naming new features
1973	The Working Group for Planetary System Nomenclature formed	
1974	NASA forms Lunar Photography and Cartography Committee	NASA attempts self-regulation over naming conventions
1979	Most of major naming problems solved	
1994	Clementine surveys moon	
1995	*Gazetteer of Planetary Nomenclature 1994*	Definitive international agreed list of named features on all planets
2007	NASA and Google set up joint venture to map moon at improved resolutions	

ATLASES OF THE MOON

After 400 years of endeavour there are now some very fine atlases of the moon. These include:

The Lunar Orbiter Photographic Atlas of the Moon
As it sounds, this is an atlas that describes the moon through the photographs and material captured in the Lunar Orbiter programme.

The Clementine Atlas of the Moon
An atlas that uses the imagery captured during the Clementine lunar fly-bys.

The Consolidated Lunar Atlas
Only 250 copies of this work were published. It is probably the most useful and most expensive lunar atlas ever published.

The Photographic Atlas of the Moon
This is the USAF *Photographic Lunar Atlas*. A wealth of material using photographs taken in various light conditions from earth-based observatories.

The Times Atlas of the Moon
The atlas is based on Lunar Aeronautical Charts produced in the 1960s. The charts were designed for astronauts and use a lot of aerial-mapping images. The atlas includes material from the Lowell Observatory.

The Greaves and Thomas Lunar Globe
Greaves and Thomas have designed and manufactured a useful and charming miniature globe of the moon. It is constructed at the same scale as their 12-inch globe of the earth. The surface of the lunar globe uses photographic data from NASA. The moon and earth globes can be used together to demonstrate lunar phenomena such as eclipses. The globes can be seen at www.globemakers.com

Chapter 3
Gods and Myths

Mythology and the Moon

The sun rises and sets and does not change. The moon rises and sets and changes every day. In its monthly cycle, endlessly repeated, the moon appears out of darkness as a silver sliver, grows into a shining round disk and then dwindles back to a sliver, before becoming invisible to us for three days. Mankind has seen in this a reflection of its own mortality. The moon, like man, is born, grows to full splendour, fades, and then dies.

This monthly cycle has exercised a pull on the imagination of man no less strong than the lunar gravity that affects our tides. We have used

the moon as a clock to count our days, and as a god to count our lives. For thousands of years a confused, rich and ever-changing mythology has grown around the moon. This mythology has always attempted to explain what is happening in the sky. It often links the sun and the moon, making them lovers, brother and sister, father and child, wife and husband. Our ancestors gave the moon a personality, ascribed to it powers and desires and worshipped it. They attempted to placate and cajole it

A late-fifteenth-century representation of the sun and the moon.

to their own ends, to draw down its power.

In the West the moon has become a beneficial feminine power, a natural goddess. This was not always so. One of the earliest lunar gods was Sin, worshipped 5000 years ago in the Mesopotamian city of Ur. The very name Allah (al-Llah – the God) may be derived from the moon god Hubal. In Germany, the sun is female and the moon male – Frau Sonne and Herr Mond. Many Sanskrit names for the moon are male, including Kandra, Soma and Vidhu.

As Western civilization passed from Mesopotamia to Ancient Egypt, from Ancient Greece to Rome, and from the Middle East to Europe, mythology and religion rolled with it like the tide, each new version carrying odd bits of the old.

The Mesopotamian moon god Sin gave way to the Egyptian moon god Thoth, 'the god of gods'. Thoth traversed the skies with his companion Re, the sun. Thoth assisted at the reckoning of time, oversaw wisdom and learning, and recorded the weights of the hearts of the dead. If you angered him he would cut your heart out, too. He was so important that huge statues weighing 30 tons each were erected to him in the temple at Hermopolis.

In its turn, Egypt faded as a great civilization and was replaced by Greece, whose ideas were to reverberate through the millennia down to our own times. Greek mythology feminized the moon deities, giving us Artemis and her associates Selene and Hecate, who later, under the Romans, became Diana, Luna and the less well-known Trivia. There are still Diana cults in Europe today.

Other cultures had other myths. In India the Hindu moon was a chariot to transport the Hindu Moon Lord Soma through the twenty-seven mansions of his monthly transit. Soma in his turn was replaced by Chandra, the controller of life from birth to death.

Many religious festivals and myths are linked to the moon's phases. The first Chinese calendar, Hsiu, was lunar. Twenty-eight warriors protected the moon on its journey, one for each phase. In Mesopotamia, the Sumerians had a festival, Sabbatu, to implore Sin to return to his former state. This is the source of the Hebrew Sabbath. The calendar used by the Egyptians was lunar; it became hopelessly unwieldy. The Christian festival of Easter is calculated by the cycles of the moon. Christians believe that Christ died, was

Mesopotamian seal (detail) showing the symbol of the god Sin.

buried, descended into hell and three days later rose again, having vanquished death. There are similar stories all over the world of a hero figure or god descending into a hellish abyss, conquering death and triumphantly returning three days later.

Every month the moon vanishes for three days before it slowly reappears. This celestial rhythm is seen and felt by every creature that has ever lived on the earth. Mankind has invented many stories to explain the mysteries of the moon. These stories have been told for thousands of years. They have been recited, performed and read by moon, fire, candle and electric light. They have been heard in caves, temples and halls all over the planet. The stories are full of confusion, fear, despair and hope. They contain unimaginable people and events: heroes, monsters, fiends and ghosts. The power they have always had over us can still be felt today, and will probably continue until the moon stops rising and we cease to exist.

MOON GODS AND GODDESSES

These dates are a rough guide to the life of the country or culture. The god concerned may have existed for all or part of the period and may have had several forms.

God	Culture/Country of origin	Approximate date range	Description
A (Sirdu)	Mesopotamia	3000–550 BC	Moon goddess
Alignak	Inuit	AD 1000–present day	Moon and weather god
Anna Perenna	Rome	500 BC–AD 500	Goddess of new year
Annit	Mesopotamia	3000–550 BC	Original ruler of moon, superseded by Ishtar
Artemis (Cynthia/Delia)	Greece	1500–30 BC	Goddess of the hunt, chastity and childbirth
Artimpassa	Mesopotamia	3000–550 BC	Goddess of love and the moon
Ashima	Mesopotamia	3000–550 BC century BC	Moon goddess

Athenesic	Native N. American	AD 500–1500	Moon goddess
Bendis	Greece	1500–30 BC	Moon goddess
Bong	India (Hindu)	2000 BC– present day	Devoted sister to Bomong, moon goddess
Candi	India (Hindu)	2000 BC– present day	Counterpart to Chandra
Chandra	India (Hindu)	2000 BC– present day	His mother swallowed the moon
Chang'e	China	1200 BC– present day	Goddess. Floated into exile on moon
Chang'o	China	1200 BC– present day	Goddess. Another name for Chang'e
Chup Kamui	Japan	AD 100– present day	Sun goddess who started as moon goddess
Coatlicue	Aztec	AD 500–1500	Terrible mother of all the gods
Coyolxauhqui	Aztec	AD 500–1500	Her severed head became the moon
Cynthia (Artemis/Delia)	Greece	1500–30 BC	Goddess of the hunt
Delia (Cynthia/ Artemis)	Greece	1500–30 BC	Goddess of the hunt
Diana	Rome	500 BC– AD 500	Goddess of wild animals and the hunt
Fati	Polynesia (Society Islands)	1600 BC– present day	Personification of the moon
Gnatoo	Polynesia (Friendly Islands)	1600 BC– present day	Goddess
God D/Itzamna	Mayan	AD 500–1500	Rules night
Hanwi	Native N. American (Oglala tribe)	AD 500–1500	Shamed into being moon
Hecate	Greece	1500–30 BC	Goddess of crossroads
Hina/Tapa	Polynesia	1600 BC– present day	Goddess. Creator of first coconuts
Hina-i-ka- malama	Polynesia (Hawaii)	1600 BC– present day	Goddess seen on face of moon

»

God	Culture/Country of origin	Approximate date range	Description
Hubal	Pre-Islamic	pre-AD 600	God. Possible forerunner of Allah
Huitaca	Native S. American	AD 500–1500	Goddess of drunkenness and pleasure
Hunthaca	Native S. American	AD 500–1500	Moon goddess
Ina (U)	Polynesia	1600 BC–present day	Woman who works in the moon
Ishtar	Mesopotamia	3000–550 BC	Goddess. Shed clothes on trip to underworld and moon darkened
Isis	Egypt	3000–51 BC	Goddess. Patron of nature and magic
Ixchel	Mayan	AD 500–1500	Goddess of healing, procreation and floods
Ix Huyne	Mayan	AD 500–1500	Moon goddess
Jacy	Native S. American	AD 500–1500	Guards moon
Juno	Rome	500 BC–AD 500	Goddess of new moon
Ka-Ata-Killa	Native S. American	AD 500–1500	Moon goddess
Khons	Egypt	3000–51 BC	God. The Wanderer and full moon
Luna	Rome	500 BC–AD 500	Moon goddess
Mah	Persia	550–350 BC	Makes plants grow
Mamma Quilla	Inca	AD 500–1500	Mother moon protects marriage and menstruation cycle
Mani	Norse	1000 BC–AD 800	Moon god
Mawu	Africa (Fou people)	AD 100–present day	Goddess of joy and fertility
Metzli	Aztec	AD 500–1500	Moon goddess (female form of Tecuciztecatl)
Nanna	Mesopotamia	3000–550 BC	God of full moon

Pandia	Greece	1500–30 BC	Moon goddess
Phoebe	Greece	1500–30 BC	Goddess. Name means brightness
Rabie	Indonesia	AD 700–present day	Moon goddess
Sardnuna	Mesopotamia	3000–550 BC	Goddess of new moon
Selene	Greece	1500–30 BC	Goddess. Rides charriot
Sin	Mesopotamia	3000–550 BC	Most important Mesopotamian moon god
Sinag	Philippines	1600 BC–present day	Moon god
Sirdu (A)	Mesopotamia	3000–550 BC	Moon goddess
Soma	India (Hindu)	2000 BC–present day	Cursed to weaken, die and revive every month
Tanit	Phoenicia	1500–300 BC	Has strange symbol
Tecuciztecatl	Native Middle American	1000 BC–present day	Moon god (male form of Metzli)
Thoth	Egypt	3000–51 BC	God. Scribe of gods, weigher of hearts
Tiazolteotl	Aztec	AD 500–1500	Goddess of lust and sex
Titania	Rome	500 BC–AD 500	Another name for Diana
Tsukuyomi	Japan	AD 100–present day	Moon god despised by his sister, the sun goddess Amaterasu
Ursula	Slavic	AD 600–present day	Also known as horse
Yellow Woman	Native N. American	AD 500–1500	Moon goddess
Yolkai Estasan	Native N. American (Navajo)	AD 500–1500	White shell woman, creates fire and maize
Zarpandit	Mesopotamia	3000–550 BC	Goddess worshipped nightly when moon appears
Zirna	Etruscan	800–400 BC	Moon goddess

Moon Gods and Their Stories

AFRICA

 Mawu
Goddess

Mawu is the most powerful deity of the Fon people in Dahomey and West Nigeria, west Africa. Her domains are joy and fertility. She brings the night and cool air, which represent wisdom and age. She has a twin brother, Liza, God of the Sun. Liza brings the day and represents heat and strength. Mawu and Liza have a son, Gu, who is a blacksmith. Mawu and Liza created the universe helped by their son and by the cosmic snake Da. An eclipse of the sun is a sign that Mawu and Liza are making love.

AMERICA

Native American is a term used to describe the indigenous population of the American supercontinent who lived there before the arrival of Columbus in October 1492. The indigenous peoples had started to colonize the continents up to 50,000 years ago. Native North American describes any culture on the North American continent; Native Middle American describes any culture, including Mayans and Aztecs, broadly occupying the area covered by Mexico and Central America. Native South American refers to Incans and any cultures or civilizations inhabiting the continent of South America.

Native North American

 Yolkai Estasan
Goddess

Yolkai held sway over the seas and the rising sun. Made of abalone, she was called the White Shell Woman. She created fire and maize from the abalone shell and was the sister of the sky goddess Estsatlehi.

Hanwi

Goddess

Hanwi lived with the sun god Wi until she was tricked into giving up her seat next to him. Shamed, she left Wi's home to go her own way. Her punishment was to give up rulership of the dawn and to hide her face when she was near the sun.

Native South American

Huitaca

Goddess

Huitaca is the goddess of pleasure and drunkenness for the Native South American Chibicha people. She is depicted as an owl and is also called Chia. She fights all the time with her husband Bochica, who represents steadfastness, sobriety and hard work. She unleashed a destructive and violent flood. As a punishment Bochica threw her into the sky, where she became the moon.

Jacy

God

Jacy guards the moon and is the creator of plants.

Ka-Ata-Killa

Goddess

Protector of the Auchimalgen tribe.

Inca

Mamma Quilla

Goddess

Mamma Quilla was the third most important Inca deity after Inti, god of the sun, and Illpapa, god of thunder. Her name means Mother Moon. She protects women, marriage and the menstrual cycle. Her legends say that she cries tears of silver and that when she is eclipsed she is fighting animals that are attacking her. If the animals succeeded in killing her there would be perpetual night. During an eclipse, people try to frighten the animals away by shouting and making loud noises with drums. Another legend says that the spots on the moon were caused by a fox who fell in love with Mamma Quilla. He tried to live with her but she squeezed him to her and his blood formed the dark patches which we call seas. She had her own temple at Cuzco, where one of her images was a huge silver plate, bigger than a man, which covered a whole wall.

Native Middle American

Aztec

 Coyolxauhqui
Goddess

Coyolxauhqui was the daughter of Coatlicue, the terrible mother of all the gods. Coatlicue gave birth to the moon, the sun and the stars. She was a very frightening creature, with a bosom covered with the relics of her children, skulls and bones. She represented life and death, as she has within her both a womb and the grave. She was impregnated by a ball of feathers. This disgusted Coyolxauhqui, who persuaded Coatlicue's 400 sons to slaughter her as punishment. Having killed her, they cut off her arms and legs and decapitated her. As she lay dying, her son Huitzilopochtli sprang armed from her womb and killed the 400 children, including Coyolxauhqui. He cut off Coyolxauhqui's head and threw it into the sky, where it became the moon. He then threw the arms and legs of his dead brothers into the sky, where they became the stars. And there they remain, where their mother can see them every night.

Coatlicue.

Tecuciztecatl
God

Around the time of the Spanish Conquest the Aztec gods were in despair and in need of a new sun. They called for volunteers to throw themselves into a sacrificial fire and so become the sun. Two gods stepped forward: the rich and powerful Tecuciztecatl, Lord of the Conch; and the poor, feeble goddess Nanhuatzin. Being powerful and rich, Tecuciztecatl was given the honour of flinging himself into the flames. Four times he tried and four times his courage failed. It was then Nanhuatzin's turn to try. She flung herself into the inferno, from which she rose as a magnificent sun. Ashamed of his cowardice, Tecuciztecatl threw himself into the dying embers and rose up as the moon. The other gods were angered by this and threw a rabbit at him to hide his brilliance and hurt his face. The rabbit is still there and can be seen on the face of the full moon.

Tiazolteotl (Tiaculteutl)
Goddess

Tiazolteotl is the goddess of lust and sex. Her name means Lady of the Dirt. She encouraged and forgave those who lusted after and took part in illicit sex. Legend has it that she appears to dying men and eats their sins. Her image may correspond to the four main phases of the moon. She is also shown carrying or riding a broomstick.

Mayan
Ixchel
Goddess

Ixchel is associated with healing, procreation, floods and rains, and weaving. She is referred to as 'Our Grandmother'. She used to be thought of as the archetypal moon goddess. However, recent speculation links her more with the waning moon. One legend has it that Ixchel and the sun were lovers. The jealous sun would often tell her to leave heaven, only to set off to find her again. But Ixchel would elude him by travelling only in the night sky.

Itzamna
God

Itzamna brought culture to the Mayan people. He is the moon god who rules the night, and is referred to as the Lord of Knowledge. Ixchel is his mother and his symbol is a snake. He is shown as an old man with sunken cheeks.

CHINA
1200 BC–PRESENT DAY

China is the oldest continuing civilization in the world. Its history can be traced back over 6,000 years. Paper, printing and gunpowder were all invented in ancient China, as was the use of the compass. The Chinese calendar is luni-solar and many important Chinese festivals, including the New Year, are linked to the moon.

Chang'e
Goddess (Chang'o /Chang Ngo)

There are several myths about Chang'e. One is that she and her husband, the skilled archer Houyi, were immortals living in the sky under the Jade Emperor. The Jade Emperor had ten sons who transformed themselves into suns. Their heat scorched the earth and threatened to destroy it, so the emperor asked Houyi for help. Houyi used his skills to shoot down all but one of the emperor's fiery sons. But then the emperor was angry, mourning his loss, so he banished Houyi and Chang'e to earth to live as mortals.

The Jade Emperor had ten sons who transformed themselves into suns.

Chang'e was heartbroken at her loss and Houyi went on a quest to find the secret of immortality to give to Chang'e. He found it in the form of a pill, which he brought home and kept in a box, telling Chang'e not to look inside. Houyi went away, leaving Chang'e on her own. She was fascinated by the box and could not resist looking in it. Nor could she resist the deliciously fragrant pill. She put it in her mouth and immediately began to float. At that instant Houyi came home and saw his wife floating into the sky. He could not stop her, and she eventually landed on the cold, bright, lonely moon. She still lives there with two companions: a jade rabbit who can be seen in the moon's dark spots; and a woodcutter named Wu Gang. Wu Gang was banished to the moon and told that he can only leave when he has cut down a cinnamon tree that grows there. Every time he cuts the tree down it instantly grows back. Wu Gang will work on the moon for ever.

China's first moon probe was called Chang'e 1.

Chang'e flying to the moon.

EGYPT
3000–51 BC

The Ancient Egyptians flourished for 3000 years in the valley of the Nile. We know the names of 1500 of their deities, and their stories form a complex and sophisticated web of myth and superstition. The two lunar deities were Thoth and Khons.

Thoth
God

Thoth is the god of the moon and wisdom. His images are to be found in sculpture, stone reliefs and wall paintings from 3000 BC to the end of Egyptian history in AD 400. Writing about him can be found in pyramid texts and coffin texts. He was born from the head of the god Seth. He is depicted variously as part human, part ibis; all ibis; or as a seated baboon. He wears a crown of a crescent moon surmounted by a moon disc. Generally benign, as the scribe of the gods he is responsible for entering the record of the souls who pass into the afterlife. He is the inventor of arts and science and the master of magic. If angered, he will decapitate the adversaries of truth and tear out their hearts.

Khons
God

Khons is also known as the Wanderer. The moon god Khons was recognized from 2500 BC. He was an important god at Thebes, where he is described as being the son of Amun and Mut. His sacred animal is the baboon. He wears a crown consisting of a crescent moon supporting a full moon orb.

Isis
Goddess

Isis is the patron of nature and magic and a friend of the underdog. She is the ideal wife and is described in the Book of the Dead as: 'The Moon Shining over the Seas', 'The Brilliant One in the Sky' and 'The Bringer of Light to Heaven'.

ETRUSCAN

Etruscan is the word used to describe the peoples of ancient Italy and Corsica whose civilization flourished from about 800 BC until it was assimilated into the Roman Empire in about 400 BC.

Zirna
Goddess

Goddess depicted wearing a half-moon round her neck.

GREECE
1500–30 BC

The long cultural history of Ancient Greece started before 1500 BC and continued until about 30 BC, when Greece became part of the Roman Empire. Greek ideas in philosophy, culture and politics had a profound influence on Western culture that exists to the present day. The Romans inherited Greek mythology and modified it. The Greeks had several goddesses associated with the moon who supplanted each other with time. Their names changed but their characteristics were very similar.

Artemis (Cynthia/Delia)
Goddess

Artemis, the Greek goddess of the hunt, became Diana to the Romans. She is traditionally associated with the moon but in classical times was never shown as a moon goddess. She was described as suckling the young of all wild things. She could be both caring and cruel. She presided over the initiation ceremonies of young girls and yet, as goddess of chastity, she was threatening to young women who wanted to be married. Her temple at Ephesus was one of the Seven Wonders of the World.

Bendis
Goddess

Bendis was very similar in style to Artemis and therefore linked with the moon. She was often shown as a huntress.

Hecate
Goddess

Hecate was a strange and spooky goddess. Her sacred places were graveyards and three-pronged crossroads, where sacrifices could be left for her. The Romans adopted her with the name Trivia – three roads. She is shown with three heads: a dog, a snake and a horse. Two ghost hounds followed her. She has been adopted by neo-pagans as the patron of witchcraft and evil, and her plants included hazel, black poplar and willow.

Hecate.

Pandia
Goddess

Pandia was Selene's daughter. She was the goddess of the full moon, called the Utterly Shining Full Moon. Her father was Zeus.

Phoebe
Goddess

Phoebe's name implies brightness. She has become associated with the moon and she is really the same as Artemis.

Selene
Goddess

Selene was beautiful, glamorous and romantic. She was the goddess of the moon and was shown as a pale-faced woman, crowned by a full moon, who rode a silver chariot pulled by white horses or oxen. Her brother was the sun god, Helios. Helios passed through the sky during the day, and as night fell Selene followed him to meet her sister Eos, goddess of the dawn. The Romans knew her as Luna, which is the Latin for moon. The study of the geology of the moon is called selenology.

INDIA
2000 BC–PRESENT DAY

Civilizations have existed in India since about 3300 BC. The Buddhist, Hindu, Jain and Sikh religions originated there. Many Indian festivals are linked to the moon, sun and stars. The festivals called 'chaturti' and 'ekadashi' take place on the day after the full or new moon (the 'tithi'). The festivals of Deepavali and Navaratari are based on the new moon of a particular month.

Chandra
God

Chandra was young and dashing. He rode his chariot, the moon, through the night sky pulled by ten white horses or by an antelope. He carried a club and a lotus flower. Chandra was born after his mother swallowed the moon. Chandra married twenty-seven

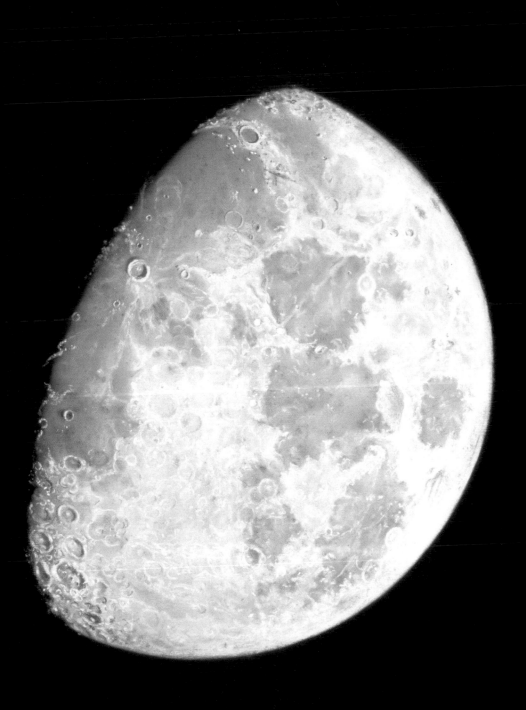

The moon has great power to fascinate and inspire. In the late eighteenth century art and science combined in this near-perfect photographic pastel drawing by John Russell. The work was based on twenty years of observation and took four years to render.

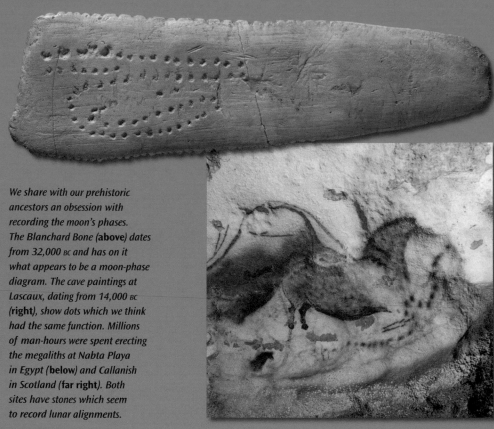

We share with our prehistoric ancestors an obsession with recording the moon's phases. The Blanchard Bone (**above**) dates from 32,000 BC and has on it what appears to be a moon-phase diagram. The cave paintings at Lascaux, dating from 14,000 BC (**right**), show dots which we think had the same function. Millions of man-hours were spent erecting the megaliths at Nabta Playa in Egypt (**below**) and Callanish in Scotland (**far right**). Both sites have stones which seem to record lunar alignments.

Moon gods are not always gentle.
When angered by the adversaries of
truth, Thoth (**above**), the Egyptian
god of the moon, would decapitate
them and tear out their hearts.

Coyolxauhqui, the Aztec moon
goddess, murdered her own mother
and was in turn murdered by
her brother. She was decapitated
and her head thrown into the
sky where it became the moon.

Other moon gods are less warlike. In this eleventh-century English manuscript, Diana (**above left**), the Roman moon goddess, is shown riding through the sky in a chariot. The Greek moon goddess Selene (**left**) fell in love with Endymion and tried to keep him young for ever. The Inuit moon mask (**above**) has feathers to represent the heavens and would have been used in ceremonial dances.

The Roman empire vanished but the torch of astronomical thought was preserved in the world of Islam. This miniature (**above**) is an eighteenth-century copy showing a thirteenth-century Armenian astronomer at work. At his feet can be seen the instruments that had remained unchanged for three thousand years and which used the same principles as the giant instruments of the eighteenth-century observatory at Jantar Mantar in India (**opposite, bottom**). Confusingly, the Armenian astronomer is using a telescope, perhaps added by the eighteenth-century copyist.

The telescope was available to seventeenth-century and Ottoman astronomers (**opposite, top**), who seem to be using optical instruments to work out the relationship between the earth and the moon. This relationship is neatly shown in the sixteenth-century Iranian illustration of the moon's phases (**opposite, top right**). Galileo's famous drawings (**left**) are among the first to be made with a telescope.

The moon's relationship with the weather has caused intense speculation for thousands of years. After the Enlightenment of the eighteenth century, scientists tried to discover a rational link. They were not always successful. The weather table (**right**), based on speculation by John Herschel, caused him considerable embarrassment. The picture above is an artist's impression of the moon, the sun, and meteorological phenomena.

daughters of Daksha and neglected all but one, so Daksha cursed him to die of consumption. But as he weakened, so did all things on earth. Daksha mitigated the curse, turning it instead into a monthly waxing and waning. Chandra was father to a race of lunar kings. He was also the protector of rabbits, which were sacred to him. He is said to have built a temple of gold to Lord Shiva, Master of the Moon. He has another incarnation as the god Soma.

Candi
Goddess

Candi was the goddess counterpart of Chandra, Lord of the Moon. They took it in monthly turns to be the moon.

Soma
God

Soma represents a sacred yellow drink, soma, that is thought to have divine powers. He is Lord of stars, plants and Brahmans. Soma presides over Mondays.

Bong and Bomong
Goddesses

The sisters Bong and Bomong were sun goddesses and devoted twins. When the creatures of the earth realized they could not live with two suns, each one shining for twelve hours every day, they decided to kill Bong. The plan backfired when Bomong, grieving for her sister and fearing for her own life, went into hiding, leaving the earth in perpetual darkness. The earth beings, realizing they needed the balance of both night and day, re-created Bong as the moon, the light of the night sky.

INUIT
ABOUT AD 1000–PRESENT DAY

The Inuit are a group of peoples who live within the Arctic Circle in Alaska, Greenland and Canada. Their extreme climate has shaped their belief system. Inuit beliefs are more a narrative about the world

Inuit depiction of the spirit of the moon, 1893. The face symbolizes the air, the hoops the cosmos and the feathers the stars.

and the people that inhabit it. An Inuit shaman once described his world as being one in which 'we don't believe, we fear'.

Alignak
God

Alignak is god of the moon and the weather. He is responsible for eclipses of the sun and moon, storms, earthquakes and other natural phenomena.

JAPAN
FIRST CENTURY AD 100–PRESENT DAY

Japanese mythology derives from Shinto and Buddhism. A belief that every natural thing has a spirit has led to an enormous cast list of deities and spirits.

Chup Kamui
Goddess

Chup Kamui is a sun goddess, though she started life as a moon goddess. She found it too painful to look on the adultery that took place in the world at night and implored her brother, the sun, to change places with her. He agreed.

Tsukuyomi
God

Tsukuyomi is the god of the moon in Japanese Shinto mythology. His father is called Izanagi and in mythology is the creator of the first world. After returning from the underworld, where he had gone to seek his wife, Izanagi bathed to purify his sins. As he washed his right eye, Tsukuyomi was born. Tsukuyomi has a sister, Amaterasu, who is a sun goddess. Amaterasu was born from Izanagi's left eye. The brother and sister climbed a celestial ladder to live in the heavens.

But Tsukuyomi and Amaterasu are sworn enemies. Amaterasu sent Tsukuyomi to represent her at a feast given by the food god, Uke Mochi. Uke Mochi had prepared a beautiful banquet by disgusting means. She faced the sea and spat fish out of her mouth. She faced the woods and spat out game and then she faced the paddy fields and spat out rice. This so enraged Tsukuyomi that he killed his hostess. When Amaterasu heard what had happened she swore she would never look at her brother again. This is why day and night are never in each other's company. In Japanese, *tsuki* means moon and *yomi* means counter of the months.

MESOPOTAMIA
ABOUT 3000–550 BC

The area known as Mesopotamia (roughly modern Iraq) saw the development of the earliest civilization that has left written records. Its time of greatest power was from 3000 to 1500 BC. Mesopotamia is situated between the Tigris and the Euphrates and its name means 'the land between two rivers'. It is sometimes referred to as 'the cradle of civilization'. The Sumerian, Akkadian, Babylonian and Assyrian empires were centred in Mesopotamia. As in many cultures, their moon mythology is complex and confusing. The most important lunar deities are Nanna and Sin.

Sin (Nanna)
God

Sin was the most important moon god throughout the life of the civilizations of Mesopotamia. Sin can refer to the crescent moon and Nanna can refer to the full moon. Sin's name came to mean Chief God and the Father of all the Gods. He was depicted as a wise old man, with a flowing blue beard, who rides on a bull. The word Nanna means 'Illuminator'. Sin's main temples were at Ur and Harra. The temple at Ur was called the House of the Great Light. Sin's number was 30, which may refer to the number of days in the lunar month (29.53). His symbols are the crescent moon, the bull and the tripod. He was the father of the sun god Shamash and the goddess Ishtar, who represents the planet Venus and fertility.

Zarpandit (Zerbanit, Zerbanitu, Zerpanitum, Beltis)
Goddess

Zarpandit was a goddess worshipped nightly at the appearance of the moon.

Ishtar (Ashdar, Astar, Istar, Istaru)
Goddess

Ishtar was derived in part from the goddess Innana. She was daughter of the moon god Sin, and sister of Shamash, the sun god. On a trip to the underworld to find her dead lover, Tammuz, she had to shed her clothes, which caused the moon to darken. On her return she regained her clothes and the moon brightened once more.

Annit
Goddess

Annit was superseded by Ishtar. Originally the ruler of the moon, she was portrayed as a disc with eight rays. Annit and the moon god Sin were known to help mortals.

PERSIA
550–350 BC

This period describes the first great Persian empire, which covered most of Greater Iran – over 7 million square miles. Much of its religious and philosophical thought was based on the teachings of the prophet Zoroaster and is known as Zoroastrianism.

Mah
Goddess

Mah was a minor Persian moon goddess. Her light was believed to make plants grow. Her name means moon and is a very common girl's name in Iran and Iraq.

PHOENICIA
1500–300 BC

The Phoenicians were powerful maritime traders and their civilization had a huge impact on the countries of the Mediterranean. The Phoenician alphabet is thought to be the ancestor of modern alphabets.

Tanit
Goddess

Tanit was a lunar goddess of war, a nurse and a virginal mother. It is thought that child sacrifice formed part of her ritual. In Egyptian, her name means Land of Neith and Neith was a war goddess.

The symbol for Tanit.

POLYNESIA
1600 BC–PRESENT DAY

Polynesia describes a group of about a thousand islands in the Pacific Ocean. Cultures have existed on the islands from about 1600 BC and they share a remarkably consistent mythology about the creation of the earth, the sky, the stars and the ruling gods. Polynesians were renowned for their navigational skills and would travel across thousands of miles of ocean to attend religious festivals.

Gnatoo
Goddess

Gnatoo is the moon goddess of the Friendly Islands. Her portrayal, as a woman pounding out tapa (a cloth made from beaten bark), is typical of Polynesian woman-in-the-moon myths.

🌝 Hina (Tapa)
Goddess

Hina is the goddess of all goddesses. Her name means 'woman who works in the moon'. In Hawaiian mythology her full name is Hina-i-ka-malama. There are many myths and stories surrounding her. In one story she sailed her canoe to the moon. In another she was making a noise that angered her brother, who threw her into the heavens. In Tahitian and Hawaiian myths, she escaped the drudgery of beating tapa by fleeing to the moon. In another, a chief lured her up from a land under the seas and her gourd became the moon. Another myth credits her with creating the first coconuts with Te Tuna – the eel god.

🌝 Hina-i-ka-malama
Hina in the moon

After beating cloth from tapa bark day after day, Hina decided to take the 'rainbow path' to rest on the sun. The sun was too hot for her so she climbed down to the moon. She can be seen when the moon is full.

🌝 Fati

Fati embodies the moon on the Society Islands.

🌝 Sina (Ina/Hina-Ika)
Goddess

Sina is a moon goddess. She kept an eel in a jar, but when it grew as long as a man she let it swim free in a pond. One day, while she was bathing in that same pond, the eel assaulted her. She cried out for help and was saved by the people of Upolo, who sentenced the eel man to death. Before dying, the eel man (who was the god Te Tuna) asked Sina to bury his head in the sand on the seashore. She followed his request and after some time the first coconut palm grew there; a gift from the gods.

This refers to the history of the Arab peninsula before the rise of
Islam in about AD 630. There is not much archaeological evidence
and written sources are generally Arabic transcriptions of an oral
tradition or translations made by later cultures.

Hubal
God

A moon god and the most powerful of the 360 gods worshipped at
Mecca in pre-Islamic times. It is believed that Hubal may have been
a forerunner of al-Llah, which means 'The God', and may be the
reason that mosques have a moon sign over them. Hubal had three
daughters. The devil tricked Muhammad into saying in the Koran
that they should be worshipped. The lines that Muhammad wrote in
the Koran about the daughters are known as the Satanic Verses.

ROME
500 BC–AD 500

The Romans inherited a lot of their mythology from the Greeks.
They changed names and added attributes from other similar gods
they encountered as their empire grew.

Anna Perenna
Goddess

Anna Perenna was the Roman goddess of the New Year and
renewal. Her name means 'eternal stream' and has associations
of the completed year going on for ever (annual and perennial).
Her festival was celebrated around the Ides of March, the 15th of the
month. In March the winter and the old year have ended, spring and
growth have started, and with them the cycle of life begins again.

Diana
Goddess

Diana is the moon goddess. Her ancestry lies with the Greek moon
goddess Artemis, and she replaced Luna, an earlier Roman

moon goddess. Diana is a huntress and is seen carrying arrows, accompanied by wild animals and hunting dogs. She is a chaste goddess of fertility.

It is best not to anger her. Her legend tells of the fate of the handsome young hunter Actaeon. One day while out hunting with his hounds, he came across Diana bathing naked. Knowing that he had seen her she became very angry and turned him into a stag. His own hounds chased him and tore him to pieces. Later mythology has Diana, Luna and Hecate as three elements of the same spirit. Luna rules the heavens, Diana the earth and Hecate the underworld.

Juno
Goddess

Juno is called Queen of the Heavens. Her sacred day was the *Kalends*, in the Roman calendar the first of each month and the time of the new moon. She is associated with all the cycles of womanhood. She is depicted as warlike in dress and wears the goatskin cloak of the soldier. She is often armed.

Luna
Goddess

Luna is the Roman version of the Greek goddess Selene. Like Selene she is beautiful, glamorous and romantic. She is crowned by a full moon and rides a silver chariot pulled by white horses or oxen. Her role as moon goddess was usurped by Diana.

Titania
Goddess

Another name for Diana.

Trivia
Goddess

Trivia is the same as the Greek goddess Hecate. Her name means Goddess of the Three Ways. She held sway over crossroads, which were dangerous places at night. Statues were erected to her at road junctions and feasts were held there under the full moon. She was known as the Queen of Ghosts.

Moon Myths

Myths are traditional stories that attempt to explain natural or social phenomena. There are many myths about the moon, what it is, how it was formed and what the markings on its surface mean.

In a common European folk tale, a couple who are found working on Christmas Eve are punished by being separated for all time. They are given a choice: freeze on the moon or burn on the sun. The woman chooses the sun. The man chooses the moon, where he can still be seen today with his dog and his lantern. Where Western cultures see a man in the moon other cultures see a rabbit, a hare, a frog, a woman or even the Yin and Yang symbol.

A sixteenth-century depiction of the rabbit in the moon.

CHINA

P'an Ku creates the universe

At first there was nothing, only chaos. Then P'an Ku was born, the child of Yin and Yang, the dual powers of nature. For 18,000 years he created the universe: the sky, the moon, the stars and the earth. But he forgot to set the sun and the moon in the sky. Instead, they went to the Han Sea, leaving the world in darkness. The Steward of Time was sent by the Earthly Emperor to make the sun and moon move through the heavens but they refused. In the end Buddha intervened. He ordered P'an Ku to write the character of the sun on his left hand and the character of the moon on his right. He was then instructed to go to the Han Sea and raise his right hand to call the sun and his left to call the moon. He performed the ritual seven times. The sun and the moon took their places in the sky, and divided the darkness into day and night.

Yue-Lao, the old man in the moon

Yue-Lao is responsible for deciding mortal marriages. He ties future husbands and wives with a magic silk thread which cannot be broken until death and which draws them together into marriage.

ENGLAND

When the moon was kidnapped

Once upon a time there was a village surrounded by evil boggy land where there lived witches, demons and all manner of disgusting creatures. At night the villagers could only cross the bog when the moon was out. When there was no moon they would be attacked and even killed by the creatures. The moon decided to see for herself what was happening. She threw on a black cloak that hid her light and descended to the bog. As she walked on its slippery surface witches flew at her and hands reached up from the slime, clawing at her and trying to ensnare her. Suddenly she slipped on a stone and became trapped by brambles that curled round her. As she struggled she heard the terrified cries of a man running through the darkness of the marsh with witches chasing after him. With a desperate heave the moon threw back the hood of her cloak. The light that streamed out terrified the witches and enabled the man to

At night the villagers could only cross the bog when the moon was out.

see a safe path home. The moon was distracted as the man ran off and the creatures of the swamp overwhelmed her. She was forced into the stinking water and a huge rock was rolled on top of her.

As the dark nights went by the villagers grew more and more frightened. They wondered what had happened to the moon. The man the moon had rescued stepped forward to say that he thought he knew where the moon might be but how could they go there in the dark? The villagers approached their wise woman, who showed them how to protect themselves.

The village set out into the darkness led by the man and the wise woman. In the dark they came across the giant stone. After a great struggle they pushed it over. As it fell the light of the moon blazed out and the moon slowly rose from her captivity into the sky. The evil creatures fled and never terrorized the village again.

GERMANY

The man in the moon

A man cutting sticks on a Sunday meets a well-dressed man going to church. The churchgoing man says, 'Have you been cutting sticks on a Sunday, when all must rest?' The man replies, 'Sunday on earth, Monday in heaven. It is all the same to me.' 'Then carry your burden for ever,' says the churchgoer. 'As a warning to all Sabbath breakers, you shall stand for all eternity on the moon.' The stranger vanishes, and the man is taken up to the moon, where he can be seen to this day.

GREECE

Selene and Endymion

Endymion was a handsome shepherd who looked after his sheep on Mount Latmus. The moon goddess Selene fell in love with him when she saw him sleeping in a cave, and asked Zeus to grant him perpetual sleep so that Endymion could never leave her.

Why a rabbit can be seen on the face of the moon

The Monkey, the Otter, the Jackal and the Rabbit were friends. One evening they discussed what they would do the next day, which was a day of fasting. They agreed to be especially zealous helping others. They hoped that they would please the spirits and be rewarded for their efforts.

The next day they came across an old man sitting by a fire and begging for food. Immediately the Monkey scrambled into the trees to pick fruit and nuts for the old man. The Otter rushed to the river bank, where he found dead fish to offer the beggar. The Jackal slunk off and stole a lizard and a pot of curd. The Rabbit could only gather grass so he ate his fill and then offered his own body to the old man by jumping into the fire.

Immediately the beggar revealed his true identity and turned into the god Sakra. The Rabbit was not burned but rewarded for his selflessness. And in order that everyone should know what the

*The sun goddess
Amaterasu holding
the moon with a
hare (1407).*

Rabbit had done Sakra painted a picture of him on the face of the moon. The smoke that seems to surround the picture is the smoke from the fire that the Rabbit jumped into.

There are many variations on this story.

MAORI

Ka-Ne restores the moon

In the land above the heavens, 'the land of the water life of the gods', there is a lake called the Living Water of Ka-ne, which has the power to restore life. When the moon dies, she goes to the water to be restored to her path in the sky.

NATIVE NORTH AMERICAN

First Man and First Woman make the sun and moon

The first children of the world, First Man and First Woman, came out of the underworld and made the moon and the sun to divide the darkness. They used turquoise, red coral, and eagle, lark and flicker feathers to decorate the sun. They put the coral round his face, and used the feathers to spread the light. They decorated the moon with crystal and white shells, then attached the sun and the moon to the sky with lightning bolts. But the sun and the moon did not move, and the people of the world complained it was too hot in the East and too cold in the West. Two wise and venerable men offered to give their spirits to the discs to make them move. They attached twelve eagle feathers to the discs to guide them, just as the eagle is guided by his tail feathers. First the sun set off and completed his journey across the sky. Then the moon followed, and was helped by the boy spirit of the wind, who blew on the moon to help him. But he blew the feathers in his face and covered his eyes and the moon could not see where he was going. That is why the path of the moon through heaven wanders and is not straight like the sun's.

In another version, the first children of the earth wanted to lighten the sky and make the world brighter. Sacred things were collected by the people of the world: white and yellow pollen, rainbows and turquoise. They made the sun from clear stones, sheet lightning, turquoise and snakes. The moon was made from crystal-white shells, and his face was covered with sheet lightning and white pollen. Then the first man approached a youth to carry the sun across the sky. The youth agreed but did not want to be parted from his father, so the father followed, carrying the moon. His path was not straight like his son's, because he was older and less strong. The youth, now called Sun Bearer, stopped for three days to rest. As he did so, the moon got heavier and the father became weaker, until First Man commanded the Sun Bearer to move on.

On the fifth day, the sun stopped again. He sent his messenger Coyote (who had invented death by throwing a stone into water) to say that in order for him to move he had to be paid with a life. Very quickly, a woman died, and the sun moved on. In the night, the moon stopped and demanded the same price to move on. Quickly, a man died. Ever since, men and women have died to keep the sun and the moon moving. This is not a cause for mourning, however, but for celebration, as the dead go to join the sun and the moon in their celestial light.

NIGERIA

When water visited the sun and moon

The sun invited his great friend the water to visit. The water was delighted to accept as he wanted to meet the sun's beautiful wife, the moon. The water said to the sun, 'When I come I will have to bring all those who are with me wherever I flow. You will have to build a huge kraal for them all.' The sun readily agreed and built the kraal. The water visited and soon his presence became too much for the kraal. As more and more water filled the kraal it began to sink. To escape, the sun and the moon leapt high into the sky. They leapt so far that the sun looked no bigger than a small plum. The moon cried out, 'I warned you that he would fill our kraal to overflowing!', and those were the last words the moon ever spoke on earth.

NORSE

Norse, Scandinavian or Viking mythology was shared by northern Germanic tribes. Part of an oral tradition, the written stories date from the twelfth to the eighteenth centuries.

Odin's chariot

In the beginning Audhumla, a rich hornless cow, was formed from vapours. Audhumla lived by licking salt off stones. There was also a primeval giant called Ymir and Audhumla fed Ymir. One day a man's hair appeared on the stone. Three days later a whole man, Buri, was formed. Buri married Bestla, daughter of Bolthor, one of the frost giants. They had three sons, Odin, Vili and Ve. Odin became chief of the Aesir gods. Odin and his brothers hated the giant Ymir, so one day the three brothers killed him. They also drowned all but two of the frost giants. Then, using Ymir's body, the three brothers created the world. Ymir's bones became the mountains, his blood became the seas, and his skull became the sky. Glowing coals and sparks made the sun, moon and stars. The brothers took two fallen trees, an ash and an elm, and made a man and a woman from them. They gave them the gifts of life, knowledge and feeling. Then Odin took a son and daughter of the giants Night and Day and set them in the sky on horses.

A man called Mundilfari called his son Moon and his daughter Sun. This angered the gods, who, indignant at the man's sacrilege, kidnapped the two children and put them in the sky to guide the moon and sun. The sun leads the way and decides when the moon will wax and wane. The sun in his turn has two children in the moon, Bil and Hjuki.

There are two wolves, Hati and Skoll, who are the sons of giantesses. The two wolves spend their time chasing the sun and the moon. Hati chases the moon, and Skoll chases the sun. The wolves

Odin.

want to eat the sun and the moon. When they succeed, a terrible
time will come: Ragnarok, the twilight of the gods. When men and
women on earth see an eclipse of the sun or moon, they know that it
is the two wolves tearing chunks out. They bang drums and shout to
frighten the wolves away.

POLYNESIA (OCEANIA)

Tiwi myth

Death comes as a curse upon the human race because a mother, committing adultery with the moon, neglected her son.

Bima was married to Purukapali, the Great Ancestor. The moon man, Tjapara, met Bima and took her off into a bush to make love. She left her baby behind, and stayed too long. Purukapali came back and found the little boy dead. He sought out and killed Tjapara as revenge. Purukapali lifted his son into his arms and walked backwards into the sea, pronouncing the curse that became law: 'As my son has died and will never return so shall all men.' Purukapali never came back, but the moon Tjapara returned to the sky after three days.

RUSSIA

Siberian myth of the Sun Maiden and the crescent moon

There lived a brother and sister who had long been orphaned. The brother grows up to become a handsome man. He makes a journey across the plains to gaze up at the beautiful Sun Maiden floating high above him. The Sun Maiden sees him and is struck by his beauty. She asks the heavenly spirits if she can have him and is told, 'You have long arms, you can have whoever you want.' So the Sun Maiden takes the young man in her arms and carries him into the skies, but he is burnt by her heat. He hates being held by her and wants only to go home. Because the Sun Maiden loves him, she immediately sends him home on a white winged horse. She gives him gifts of a whetstone and a comb.

Once home he greets his sister, not knowing it is the foul witch Hossiadam, who has eaten his sister. The witch prepares to cook over the fire and, unseen by the young man, cuts a leg from the winged horse. The young man is troubled and escapes on the white horse, which now has only three legs. The witch chases him. He flings the whetstone at her. Where the whetstone falls, mountains spring up to stop the witch, but she struggles through them. He then flings the comb at her, and the comb becomes a forest to stop the witch, but this too she overcomes.

Desperate to help him, the Sun Maiden grabs the young man to pull him into the sky. The witch grabs him too and they struggle over his body and tear it in half. The Sun Maiden holds the remains of the young man. He is dying and the witch has stolen his heart. The Sun Maiden uses coal to try to make him live. It is a hopeless task. In despair she hurls his poor corpse to the other end of heaven. The young man, heartless and cold, becomes the faintly glowing crescent moon. He is only dimly visible because he reflects the light of the love of the Sun Maiden.

He floats alone, a silver crescent, cold and lifeless. He and the Sun Maiden only glimpse each other for a few short hours on the longest day of the year.

Siberian myth explaining the creation of the sun and moon

Once upon a time the sun, the moon and the earth lived happily together in the earth's garden. As time passed, the earth grew jealous of the sun and moon, so jealous that he captured them and hid them away.

This action made the gods very angry. They chose the Porcupine to rescue the captured gods. Porcupine knew there was an old custom on earth that demanded that any visitor be given food and a gift. He decided he would visit the earth.

Not suspecting a trick, the earth offered the Porcupine a meal and when he had finished asked him what he would like as a gift. The Porcupine thought for a moment and then said he would like a Mirage Horse and an Echo Spear. The earth was baffled. Where would he find such things? They did not exist. In order to fulfil his obligation he had to offer the Porcupine the most precious things he had – the captured sun and moon.

Once they were in his possession the Porcupine threw the two gods into the air, where they would float for ever, content and free from capture.

Anansi and his six sons

The spider god Anansi had six beautiful sons. Each son had a special gift. The first had marvellous eyesight and could spot trouble far off. The second could build incredible roads. The third could drink rivers dry. The fourth was an amazing hunter. The fifth could throw any stone a great distance. The sixth was so fat that he was called the Cushion. One day the six sons managed to save their father's life. This is how they did it.

Anansi was a dreamer. While walking and dreaming in the forest he fell into a river and was eaten by a giant fish. At home in the village the far-seeing son thought he could see his father in trouble. The road-building son immediately built a road so they could run to the spot where they thought their father was. When they arrived, all they could see was a river. The third son drank the river dry and there they found a fish lying on the river bed. The hunter son killed it and skinned it, releasing his father. As Anansi stepped out of the fish a giant bird flew down and scooped him up. Instantly the stone-throwing son threw a rock at the bird's beak. The bird dropped Anansi and he fell to the ground, where his fall was broken by his sixth son, the Cushion.

The owl answered that they all deserved it and that he would fly and place the light in the sky for them all to enjoy. And there it remains to this day. We can see it almost every night, and we call it the moon.

That night Anansi and his sons celebrated his rescue. Anansi wanted something to reward his sons. He saw a bright light in the forest and went to get it for their reward. As he carried the light back to the village he wondered how he would divide the light between the six sons. He turned to the wise owl for help and asked him which of his sons most deserved the light. The owl answered that they all deserved it and that he would fly and place the light in the sky for them all to enjoy. And there it remains to this day. We can see it almost every night, and we call it the moon.

Lunar Holidays and Festivals

The religious festivals of both Judaism and Islam are governed by a lunar calendar, and other major faiths have moon-related festivals, but possibly the three biggest lunar festivals in the world are the Christian festival of Easter and the Chinese festivals of the New Year and the autumn Moon Festival.

Easter is a Christian festival that celebrates the crucifixion, death and resurrection three days later of Christ. The date of Easter is subject to a complicated calculation predicated on the moon. The rule governing the calculation is this: Easter shall be observed on the Sunday after the first full moon on or after the day of the vernal equinox.

The complications of this calculation are made worse by the fact that the Eastern Orthodox churches use the Julian calendar while Western Christianity uses the Gregorian calendar. This results in there often being two dates for Easter.

Easter dates 2008–2019

	Western Christian church	Eastern Orthodox church
2008	23 March	27 April
2009	12 April	19 April
2010	4 April	4 April
2011	24 April	24 April
2012	8 April	15 April
2013	31 March	5 May
2014	20 April	20 April
2015	5 April	12 April
2016	27 March	1 May
2017	16 April	16 April
2018	1 April	8 April
2019	21 April	28 April

PAGAN LUNAR FESTIVALS IN THE WEST

Modern pagan movements have assimilated many ancient festivals into their calendars. Sometimes the dates are arbitrarily linked to an ancient deity. Sometimes they are made up at random, on the whim of the priestesses or priests of the cult in question. In the twentieth century paganism has assigned a great importance to the female 'Mother Moon'. Many female deities from antiquity have been hijacked as having lunar links, while others have just been invented.

The moon is one of the most important influences on neo-pagan thought.

Lunar festivals	
30 November	Festival of Hecate: weather
20/21 December	Winter solstice, Celtic festival of the stars: light and life
31 January	Imbolc: rejuvenation and fertility
7 February	Festival of Selene
12 February	Festival of Diana
15 March	Festival of Cybele
20 March	Festival of Isis
20 March	Festival of Oestre: fertility
31 March	Festival of all lunar goddesses
1 May	Beltane, Mayday: warmth and light
9 May	Artemis
26–31 May	Diana's Ides of May: fertility and abundance
21 June eve	Ceridwen
13 August eve	Hecate: weather and thanksgiving
September full moon eve	Festival of candles, harvest moon
31 October eve	Festival of Hecate: remembrance of ancestors

The New Year is China's biggest and most important festival.
It is celebrated all round the world, wherever there is a Chinese
community. The date of the Chinese New Year is the first day of the
first lunar month of the year. Generally this means that it will fall
on the second or third new moon before the vernal equinox. The
Chinese use a calendar system that is governed by the sun and moon,
called luni-solar.

Animals are assigned to each New Year. These animals are
similar to Western signs of the Zodiac.

Chinese New Year dates until 2019

Animal			Dates	
鼠	Rat		19 February 1996	7 February 2008
牛	Ox		7 February 1997	26 January 2009
虎	Tiger		28 January 1998	14 February 2010
兔	Rabbit		16 February 1999	3 February 2011
龍	Dragon		5 February 2000	23 January 2012
蛇	Snake		24 January 2001	10 February 2013
馬	Horse		12 February 2002	31 January 2014
羊	Ram		1 February 2003	19 February 2015
猴	Monkey		22 January 2004	8 February 2016
雞	Rooster		9 February 2005	28 January 2017
狗	Dog		29 January 2006	16 February 2018
豬	Pig		18 February 2007	5 February 2019

CHINESE AUTUMN MOON FESTIVAL

The Autumn Moon festival is the biggest Chinese festival after the New Year. It commemorates the legend of Chang'e floating to the moon, where she will stay for ever. The festival is also known as the Mooncake festival. Traditionally people eat mooncakes under the autumn moon, light fireworks and plant trees.

Dates of the Chinese Autumn Moon festival 2008–2018
14 September 2008
3 October 2009
22 September 2010
12 September 2011
30 September 2012
19 September 2013
8 September 2014
27 September 2015
15 September 2016
4 October 2017
24 September 2018

Chapter 4

Gardening and the Weather

Tides rise and tides fall, pulled by the moon's gravitational power. For thousands of years, farmers thought the same power made the sap rise and fall in their crops. The relationship of a plant to the moon was thought to be very similar to that of the human body to the moon. The argument ran that the moon has power over water. The human body contains water, therefore the moon has power over the body as well. Plants have sap. Sap is a liquid that contains water and therefore must also be subject to the pull of the moon. When the moon is waxing it will cause the sap to rise, and when the moon is waning the sap will fall. Therefore, moon gardeners believe it is better to plant when the moon is waxing or full, and to harvest when the moon is waning or dark. The waxing moon will cause sap to rise and give strength to the plant, while under a waning moon the sap is falling and the plant is ready for harvesting.

The historian and naturalist Pliny the Elder (AD 23–79) is among the first known to have written about the impact of the moon on agriculture. By the beginning of the twentieth century, a complicated system of tradition, folklore and superstition had grown around the subject. Some of the same traditions can be found all over the world. There has long been a tradition that trees should be felled when the moon is waning, as trees cut down in the full moon would be full of sap and difficult to work. The Greek historian Plutarch (AD 46–120) wrote:

> *The Moon showeth her power most evidently in those bodies which have neither sense nor lively breath. For carpenters reject the timber of trees which have been felled in the full moon as being soft and tender and subject to the work and putrefaction and that quickly by means of extensive moisture.*

Fifteen hundred years later in France, in 1669, Louis XIV made it law that trees should be felled in a waning moon between autumn and spring. The law was not changed until the French Revolution more than a hundred years later.

Similar traditions regarding the moon and the felling of trees can be found in areas as far apart as the Near East, Africa, India, Sri Lanka and Brazil.

The Swiss Federation of Technology's Department of Forestry Science carried out experiments to test the theories. They found that trees planted and held in the dark under controlled conditions showed better germination and growth if planted during a full moon. Trees felled in winter gave stronger, harder timber if felled during a new moon, though the results were very marginal.

Even scientific thinking about what the moon does and does not do to plants was rooted in tradition and hearsay, reinforced by untested oral repetition. Some of the beliefs do work, which helps give credibility to others. The simple theory of rising and falling sap has been endlessly elaborated and refined, though it is shot through with contradiction.

In the disciplines of astronomy, alchemy and medicine much of the thinking started in the ancient worlds of Greece and Rome, then made its way via Islam and the Arab countries to medieval Europe, where it flourished. Then, in about the eighteenth century, advances in empirical science gave a new understanding as to how the physical world worked. The old ideas were discredited and rejected but continued in quasi-scientific and occult forms. The same thing happened with agriculture.

For 1500 years the world believed, and was made to believe, that the earth was the centre of the universe. They believed it because Ptolemy had said it. The same Ptolemy said:

> *Farmers take into account the aspects of the Moon when at*
> *Full, in order to direct the copulation of their herd and flocks*
> *and the setting of plants or sowing of seeds; and there is not*
> *an individual who considers these general precautions as*
> *impossible or unprofitable.*
>
> Ptolemy (AD 85–165), *Tetrabiblos*

Many earlier influential thinkers had similar theories about the moon, including Pliny and Aristotle.

> *That tiny creature the ant at the Moon's conjunction keeps*
> *quiet, but at the full Moon works busily into the night.*
> *Geld hogs, steers, rams and kids when the Moon is waning.*
>
> Pliny, *Natural History*

Pliny advised that the farmer harvesting for market should harvest before the full moon, when the produce would be heaviest. On the other hand, the farmer harvesting for his own stores should do so in the waning moon as the produce would last longer.

Pliny.

Aristotle held similar views about the full moon and made observations about sea urchins to prove it:

> *This is clearly shown in the case of sea urchins, for although the roes are found in these animals directly they are born, yet they acquire a greater size than usual at the time of the full moon.*
>
> Aristotle (384–322 BC), *The History of Animals*

Aristotle's ideas about sea urchins were made all the stronger by the fact that, in the Red Sea, sea urchins do get bigger and smaller with the waxing and waning of the moon. In 1667, 2000 years after Aristotle wrote *The History of Animals*, the Royal Society in London asked this question:

> *Whether Shell Fishes that are in these parts plump and in season at the full moon and lean and out of season at the new are found to have contrary constitutions in the East Indies.*

The Society took another 250 years to gather the evidence, analyse it and answer the question. Their answer was that they weren't sure.

By the nineteenth century, steam engines dominated the industrial world. Machines powered by steam could perform repetitive actions very fast. Factories grew in size, work speeded and labour was sucked in from the agricultural communities to man the new devices. The city rather than the field became the habitat for most people. With the discovery of electricity light itself could be turned on and off at will. For the new majority, the city dwellers, cut off from their rural roots, night and day, the seasons and the so-called natural rhythms of life disappeared.

Rudolph Steiner and the Development of Biodynamics

An Austrian clairvoyant and philosopher, Rudolph Steiner (1861–1925), became very concerned that industrialized man, cut off from his rural roots, had lost his way. As a result Steiner developed a belief system known as Anthroposophy which had a big impact on twentieth-century thought. Its influence can be seen in science, art, architecture, education, medicine and farming. Steiner's beliefs were very like those of the ancient Greeks. Like them, he believed in the importance of the four elements: fire, earth, air and water. He believed that the cosmic forces surrounding the earth act on every living thing: man, animals, insects and plants. Steiner thought that the earth had to be in harmony with the cosmos. He believed that we have lost the ability to instinctively understand the spiritual and cosmic forces that surround us and that our route to happiness is to re-establish cosmic harmony. His ideas are contentious.

In 1924, one year before his death, Steiner delivered a series of eight lectures to a group of farmers. These lectures became the basis of what is known as biodynamic farming. The biodynamic farmer works to establish cosmic harmony on his land. He wants his farm to be self-sufficient and he wants the cycle of sowing, growth and harvesting to be done in accordance with the forces flowing over the earth from the rest of the universe. The biodynamic farmer uses the position of the moon, sun and planets, and the zodiac, to regulate his planting and harvesting. The moon is especially important. It is seen as an amplifier, channelling, focusing and magnifying the other cosmic forces.

Natural forces, magnified by the moon, will enhance the power of natural fertilizers like manure, blood and bone, which will promote growth and guide harvesting. But the moon will do more – it will provide spiritual nourishment. The biodynamic farmer believes that food that has been spiritually enhanced will spiritually enhance those who eat it. Through spiritual enhancement comes happiness, and from that happiness flows cosmic harmony.

Steiner's ideas on agriculture have been used and developed by biodynamic farmers and gardeners for nearly a century. They are thought by some to be nonsense but by others to be the key to the ecologically sound management, even the survival, of the planet.

SOME EXPERIMENTS TO TEST STEINER'S LUNAR THEORIES

Experiments to prove that Steiner's theories have a scientific basis are similar to those that try to prove there is a lunar effect in medicine. Quantities of data are collected and subjected to statistical analysis. Both the experiments and the conclusions suffer from inconsistent methods and results that cannot be repeated or verified. Nevertheless, there are some intriguing results.

In Germany, in 1956, Maria Thun carried out a controlled experiment with potato planting which indicated that maximum cropping occurred with that part of the crop which had been planted on 'root days' (with the moon in the optimum position for root crops).

Thun is not alone. Ulf Abel at Geissen University carried out planting experiments in which some crops (barley and oats) performed up to 20 per cent better when planted on their appropriate moon sowing days.

The effect of the moon gets stronger and weaker on a cyclical basis as the moon waxes and wanes. This is termed 'Lunar Pulse'. At the University of Paris a biochemical study into the effect of Lunar Pulse on the DNA of plants detected that those elements of DNA whose function is related to the storage of carbohydrates was stronger at the new moon, while DNA elements related to flowering was more developed at the full moon.

Professor Frank Brown stored potatoes in the dark under controlled conditions of humidity, temperature and pressure. He found that the metabolic rate of the potatoes rose and fell with the waxing and waning of the moon. Professor Brown put in over a million potato-hours to reach his findings.

One argument used against 'lunar influence' theorists is that the measurable force of the moon's gravity is so small as to be negligible. A cup of coffee held in the hand will have a greater gravitational influence than the moon. Size may not be everything, though. In 2008 a study into the earth's magnetic field found that ruminants tended to eat with their bodies aligned to its comparatively weak force. The researchers reached their conclusions from studying herds of cattle and deer visible in Google mapping photographs retrieved from all over the surface of the planet. Like the earth and its magnetic field, the moon may have a subtle strength whose power we do not properly understand.

Biodynamic farmers point to their results to prove the effectiveness of their methods. They are meticulous in the practice and zealots in the theory of their work.

THE GENERAL THEORY OF LUNAR FARMING AND GARDENING (BIODYNAMICS)

Biodynamicists believe that the moon, the sun, the planets and the signs of the zodiac all affect how and when plants should be sown, how they will germinate and how and when they should be harvested; and that the planets also influence the lives and wellbeing of animals.

The influence of the zodiac

The zodiac is the backdrop for the passage of the moon and planets through the sky. It has the twelve familiar constellations starting with Aries and ending with Pisces. Biodynamic theorists divide the twelve signs of the zodiac into four groups. Each group is connected to one of the four elements: fire, earth, air and water. The position of the moon, its rhythms, its phases and its orbit are as important to plant growth and animal husbandry as soil type, rainfall and temperature.

Just as the zodiac is divided into four parts corresponding to the four elements, so the plant is divided into four parts: seed, root, leaf and flower. These parts are linked through the four elements to the four groups of the zodiac.

The zodiac, the four elements and the four parts of the plant

	Zodiac constellation group	Element represented by constellation group	Plant part represented by constellation group
	Aries	Fire	Seed
	Leo	Fire	Seed
	Sagittarius	Fire	Seed
	Taurus	Earth	Root
	Virgo	Earth	Root
	Capricorn	Earth	Root
	Gemini	Air	Flower
	Libra	Air	Flower
	Aquarius	Air	Flower
	Cancer	Water	Leaf
	Scorpio	Water	Leaf
	Pisces	Water	Leaf

Each month the moon passes through all twelve signs of the zodiac. Each day will have a specific relationship with a part of the plant. Some days will be leaf days, others root days, others seed, and finally flower days. It is best to plant a root crop on a root day, leaf crops on a leaf day, and so on. Common sense also applies. If it is pouring with rain on the root day you plan to plant your potatoes, wait for the next root day, which will fall about nine days later.

The influence of the planets

The planets form two groups, the Inner and the Outer. The Inner Planets are those that orbit between the earth and the sun, and are connected to earthly forces. They are: the moon, Venus and Mercury. The Outer Planets are those planets in the solar system that orbit the sun beyond the earth, and are connected to astral forces. They are: Mars, Jupiter and Saturn.

The influence of the moon

Biodynamicists describe the moon as the gate between earth and heaven. The moon has many cycles. The most important of these for the gardener are: the sidereal cycle, the synodic cycle, the cycle of perigee/apogee, the nodal cycle, and the ascending and descending cycle.

Sidereal cycle
27.3 days
Sidereal means 'of the stars'. The moon orbits the earth once every 27.3 days, passing through all twelve signs of the zodiac. It is the moon's position when the seed is planted that will influence how the seed comes to fruition.

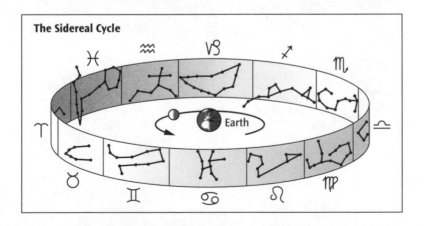

The Sidereal Cycle

Synodic cycle
29.5 days
This relates to the lunar month. The synodic cycle is the time the moon takes to pass through all its phases from new moon to first quarter, full moon, third quarter and back to new moon. It is longer than the sidereal cycle because it is governed by the moon's position relative to the sun. This cycle governs how a seed will germinate and what sort of crop it will give.

Cycle of perigee/apogee
27.2 days
The moon's orbit round the earth forms an ellipse. The point at which it is furthest from the earth is known as the apogee and the point at which it is nearest is known as the perigee. The moon's influence on water is strongest when it is nearest the earth and weakest when it is furthest away. This cycle is thought to be linked to stress in human beings and in plants. It may cause overactive sprouting, which is not necessarily a good thing.

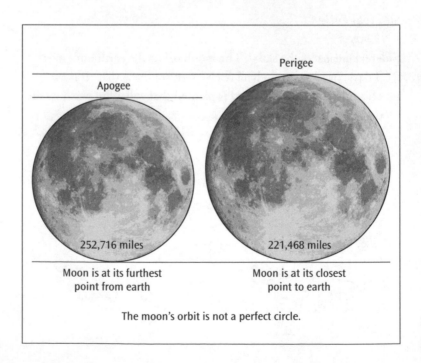

Apogee

Perigee

252,716 miles

221,468 miles

Moon is at its furthest point from earth

Moon is at its closest point to earth

The moon's orbit is not a perfect circle.

Nodal cycle
27.6 *days*

Nodal means 'crossing point'. The earth and stars orbit the sun in a fixed plane known as the ecliptic. The moon's orbit round the earth is at a slight angle to the ecliptic, which means that twice a month it crosses the ecliptic. The two points of crossing are known as nodes or 'dragon's head' and 'dragon's tail'. They coincide with the full and new moons, which are thought to have a negative effect on plant growth.

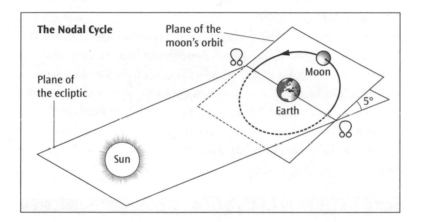

The ascending and descending cycle
27.3 *days*

The moon orbits the earth in 27.3 days. Over that period the moon follows the path of a wave. Each day it appears at a slightly higher point in the sky until it reaches a peak, after which its daily appearance is a little lower. This is the ascending and descending cycle. Biodynamic theory states that when the moon is ascending it pulls energy into the upper growing parts of a plant, and when it is descending the energy flows back down to the roots. The time taken for this cycle is the same as the sidereal cycle. This is the primary rhythm for biodynamic gardeners.

The influence of the earth

Alongside the observable cycles of the moon is the notion that the earth is 'breathing'. The rhythms of the earth's 'breathing' are daily and annual. Accordingly, biodynamicists consider the earth to have two important cycles, a daily cycle and an annual cycle.

The earth's daily cycle
The daily cycle has two phases. From 3am to 3pm the earth breathes out, and oxygen is released into the atmosphere. This is called an ascending cycle. From 3pm to 3am the earth breathes in, releasing carbon dioxide into the atmosphere. This is called a descending cycle.

The earth's annual cycle
The annual cycle is a large-scale version of the daily cycle. It also has two phases. The first marks a flow of energy out from the earth into the universe; the second is an inward flow of energy from the universe into the earth. The outward flow runs from the start of spring to the end of summer, and the inward flow runs from the start of autumn until the end of winter. Outward-flowing energy enhances growth and inward causes dormancy and hibernation among animals.

WHAT HAPPENS IN THE SOIL?

While the cosmos gives energy and spiritual food, the earth contains the physical elements needed for growth. For the biodynamic gardener, the two most important elements on the planet's surface are silica and calcium. They are essential to plant life.

▶ Plants need silica to be upright and to withstand insect attack. Silica occurs naturally as the mineral quartz. Quartz is used to make magnifying glasses, and biodynamic farmers feel they are harnessing the same powers of magnification when they use silica preparations. One of the most important preparations is a fertilizer called cowhorn silica, which is sprayed in very dilute quantities.
▶ Calcium, a major ingredient of the earth's surface, is held to work with the forces of the moon.

There are other important elements: carbon, oxygen, nitrogen, hydrogen, phosphorus and potassium.

THE INFLUENCE OF CATTLE

Cattle are very important in the biodynamic world. It is thought their hollow horns act as a antennae which capture and magnify lunar energy. This energy passes into the stomach of the cow and is mixed with what it eats. Manure from cows is highly prized, as are their horns.

The biodynamic gardener must also be sensitive to the equally important measurable and observable physical properties of his environment. These include the weather, the seasons, soil content and humidity. The biodynamic gardener tries to develop an intuitive understanding of all the forces at work in the cycle of growth and harvest. Biodynamics is a tool to aid good farming and gardening practice.

THE BIODYNAMIC THEORY IN PRACTICE

The next part of this chapter gives broad guidelines for the would-be lunar gardener, and a feel for what biodynamic gardening entails. The guidelines may seem complicated but they get easier with practice. Doing the work, watching the moon, and feeling the rhythm of the seasons are all activities that help to develop a natural understanding of what the cosmos is doing and saying. After a while the rules cease to be rules and become second nature. And remember that while all the cosmic signs may be saying 'plant now', the muddy rain-drenched garden or field may be saying, 'Wait, let it stop raining so I can get dry before you tramp all over me.' The cosmos wants things to grow and flourish. It is not an iron-willed hostile taskmaster but a mighty and benevolent force that is there to help you. Learn its rhythms, tune in to what it is saying and go with the flow.

There are several excellent handbooks and calendars available to guide the gardener through the biodynamic process. Some of the books are published annually and contain month-by-month charts showing what the moon is doing and what sign it is in on any given day. There is a list of some of these books on page 361.

There are also several very useful organizations. One of these is the Biodynamic Education Centre. One of the centre's founders is Lynette West, who is a world expert on biodynamic farming. She lectures in England and Australia and a timetable of her classes can be found on the BEC's website: biodynamiceducation.com

Planting and harvesting

The broad framework is simple. These are the steps you should follow to develop a planting and harvesting technique:

▶ Identify what you want to sow.
▶ Identify the zodiac sign that is appropriate to that plant and when in the month the moon will be passing through that sign.
▶ Organize things so you can plant in the middle of the moon's transit through the sign.
▶ Prepare the soil on the same day that you are going to sow.
▶ Sow the seeds at moonrise. This is considered to be the best time of day for planting. Moonrise when the moon is full is an especially good time to plant. The combination creates a lot of moisture, which helps the plant to germinate and to establish itself.
▶ Harvest the crop when the moon is passing through the same zodiac sign that it was planted in. Try to do this around moonset.

For example
Crop: Potato
Plant type: The potato is a root plant and should be sown on a root day when the moon is passing through an earth sign.
Sowing: Plant at moonrise when the moon is in Taurus, Capricorn or Virgo.
Harvest: Harvest at moonset when the moon is in Taurus, Capricorn or Virgo.

Biodynamic hints for agricultural activities

While you are mastering the basic technique, here are some hints to help your other gardening activities.

Transplanting

- Transplant during a descending moon (3pm to 3am). This will help plants to develop good root systems. Transplanting is also best done during a waxing moon.

Harvesting

- Harvest in the same moon sign that you planted in.
- Fruit picked during a new moon should store well.
- Fruit picked during a full moon is best eaten at once.
- Harvest during an ascending moon between 3am and 3pm. Harvest in the early morning to get the best quality from a crop. This is when the earth is breathing out, breathing oxygen and energy into the universe and plants. This is why a lettuce cut before lunch will be crisper and tastier than one cut after lunch.
- Root crops should be harvested during a descending moon (3pm to 3am). Avoid harvesting during a full moon because there will be a lot of moisture in the earth. The water content of vegetables will be high and they will not keep well.

Pruning and gelding

- These activities are best performed under a moon which is waning and descending (3pm to 3am), which causes water levels to be low.
- Avoid pruning on leaf days.
- Prune trees during a descending moon. The sap will be falling, which will encourage healing in the cut bark.

Grafting

- Grafting is best done during a waxing moon between 3am and 3pm, when it is ascending. The sap is rising in plants at this time. This is also a good time for taking softwood cuttings.
- Take hardwood cuttings during a descending moon (3pm to 3am).

Sowing

- Plant in the appropriate moon sign and as near to moonrise as possible. Avoid the hours around the nodes, perigee or an eclipse.

Weed control

▶ A very effective biodynamic weedkiller can be made by diluting with water the ashes of the seeds of the weed you want to control.

▶ The seeds should be burnt when the moon is in the sign of the weed in question. The solutions have a lot in common with those used in homeopathic medicine and are very dilute. Refer to one of the books listed at the end of this chapter to establish quantities and proportions.

Simple guide to times for burning weed seeds

Weed	Burn seed when moon is in:
Wild mustard Charlock Red dead-nettle	Aries
Ground elder Chervil Goose grass	Taurus
Wild oats Chickweed Corn grass	Gemini
Buttercup	Cancer
Dock	Leo
Thistle Coltsfoot Horsetail Bindweed	Virgo
Thistle	Libra
Black nightshade	Scorpio
Couch grass Orach	Sagittarius
Weeds that sprout at this time are rare	Capricorn
Knot grass Shepherd's purse Penny cress	Aquarius
Tufted vetch	Pisces

Manure

Apply manure during a descending moon (3pm to 3am). This includes biodynamic manure concentrates, compost and green manure crops that need to be turned in. Biodynamic gardeners use extremely dilute quantities, as can be seen from the following recipe for the fertilizer known as cowhorn silica. The recipe also illustrates the importance of the role of the cow.

To prepare cowhorn silica

Place finely powdered quartz in a cowhorn. Bury the cowhorn for the summer. While buried, the cownhorn will attract and magnify cosmic energy. In the autumn, retrieve the cowhorn. Place 0.5 grams of the quartz into 5 litres of water. Stir the solution for one hour. Stir from the outside to the centre of the bucket. This will cause a spiral which will concentrate the cosmic forces. Change the direction of stir regularly. Only one person should stir each batch. Once stirred the mixture must be used within 4 hours. Five litres is enough to spray 1000 square metres. The preparation and application should take place on the appropriate day for the crop in question: root days for root crops, and so on.

Short guide to when to do what in the garden		
Activity	**Plant type**	**Action**
Cuttings		Take cuttings in the morning and plant in the afternoon
Digging		Dig in the afternoon to keep moisture; in the morning to release moisture
Fertilizing	Green crops. Turn to enrich with liquid manure	Do this in the morning of a fruit day
Grafting		Morning on fruit days
Harvesting	Flowers	Harvest on flower days
Harvesting	Fruit	Harvest on fruit days during an ascending moon

Activity	Plant	Action
Harvesting	Grains	Harvest on seed or fruit day
Harvesting	Leaf plants	Harvest on flower days in the early morning when the moon is ascending
Harvesting	Root vegetables	Harvest on root days in the afternoon when the moon is descending
Harvesting	Winter brassicas	Harvest on flower or fruit days to maximize storage potential
Pruning trees		Prune in a waning, descending moon. Avoid full moon, when the sap will be high
Transplanting		Transplant in the afternoon

COSMIC MOMENTS TO AVOID

When planting or cultivating the soil, the times to be avoided are
when the moon, sun and planets are at points of change.
These are:

▶ The movement of the moon to the next constellation of the zodiac
  (avoid one hour before the moon passes into a new constellation).
▶ When the sun passes to a new constellation.
▶ The period each day between 12pm and 2pm.
▶ Times of lunar and planetary nodes.
▶ Twelve hours before and after an eclipse.
▶ The movement of the moon at peak apogee and perigee.
▶ Avoid harvesting during a full moon.

Guide to plant types

Plant	Element		Plant type	Zodiac sign
Apple	Fire	⬭	Fruit/seed	Aries, Leo, Sagittarius
Apricot	Fire	⬭	Fruit/seed	Aries, Leo, Sagittarius
Artichoke	Air	✺	Flower	Gemini, Libra, Aquarius
Asparagus	Earth	⅄	Root	Taurus, Virgo, Capricorn
Asparagus pea	Fire	⬭	Fruit/seed	Aries, Leo, Sagittarius
Aubergine	Fire	⬭	Fruit/seed	Aries, Leo, Sagittarius
Basil	Water	◁	Leaf	Cancer, Scorpio, Pisces
Bay	Water	◁	Leaf	Cancer, Scorpio, Pisces
Beetroot	Earth	⅄	Root	Taurus, Virgo, Capricorn
Blackberry	Fire	⬭	Fruit/seed	Aries, Leo, Sagittarius
Borage	Air	✺	Flower	Gemini, Libra, Aquarius
Broad bean	Fire	⬭	Fruit/seed	Aries, Leo, Sagittarius
Broccoli	Air	✺	Flower	Gemini, Libra, Aquarius
Brussels sprout	Water	◁	Leaf	Cancer, Scorpio, Pisces
Cabbage	Water	◁	Leaf	Cancer, Scorpio, Pisces
Carrot	Earth	⅄	Root	Taurus, Virgo, Capricorn
Cauliflower	Air	✺	Flower	Gemini, Libra, Aquarius
Celery	Water	◁	Leaf	Cancer, Scorpio, Pisces
Cherry	Fire	⬭	Fruit/seed	Aries, Leo, Sagittarius
Chicory	Water	◁	Leaf	Cancer, Scorpio, Pisces
Coriander	Water	◁	Leaf	Cancer, Scorpio, Pisces
Courgette	Fire	⬭	Fruit/seed	Aries, Leo, Sagittarius
Cress	Water	◁	Leaf	Cancer, Scorpio, Pisces
Cucumber	Fire	⬭	Fruit/seed	Aries, Leo, Sagittarius
Elderflower	Air	✺	Flower	Gemini, Libra, Aquarius
Fennel	Water	◁	Leaf	Cancer, Scorpio, Pisces
Fig	Fire	⬭	Fruit/seed	Aries, Leo, Sagittarius
Flowering plant	Air	✺	Flower	Gemini, Libra, Aquarius
French bean	Fire	⬭	Fruit/seed	Aries, Leo, Sagittarius
Garlic	Earth	⅄	Root	Taurus, Virgo, Capricorn
Gooseberry	Fire	⬭	Fruit/seed	Aries, Leo, Sagittarius
Horseradish	Earth	⅄	Root	Taurus, Virgo, Capricorn

»

Plant	Element		Plant type	Zodiac sign
Jerusalem artichoke	Earth		Root	Taurus, Virgo, Capricorn
Leek	Earth		Root	Taurus, Virgo, Capricorn
Lettuce	Water		Leaf	Cancer, Scorpio, Pisces
Marrow	Fire		Fruit/seed	Aries, Leo, Sagittarius
Mint	Water		Leaf	Cancer, Scorpio, Pisces
Mushroom	Earth		Root	Taurus, Virgo, Capricorn
Mustard and cress	Water		Leaf	Cancer, Scorpio, Pisces
Nectarine	Fire		Fruit/seed	Aries, Leo, Sagittarius
Onion	Earth		Root	Taurus, Virgo, Capricorn
Parsley	Water		Leaf	Cancer, Scorpio, Pisces
Parsnip	Earth		Root	Taurus, Virgo, Capricorn
Pea	Fire		Fruit/seed	Aries, Leo, Sagittarius
Pear	Fire		Fruit/seed	Aries, Leo, Sagittarius
Plum	Fire		Fruit/seed	Aries, Leo, Sagittarius
Potato	Earth		Root	Taurus, Virgo, Capricorn
Pumpkin	Fire		Fruit/seed	Aries, Leo, Sagittarius
Radish	Earth		Root	Taurus, Virgo, Capricorn
Rhubarb	Water		Leaf	Cancer, Scorpio, Pisces
Runner bean	Fire		Fruit/seed	Aries, Leo, Sagittarius
Sage	Water		Leaf	Cancer, Scorpio, Pisces
Sorrel	Water		Leaf	Cancer, Scorpio, Pisces
Spinach	Water		Leaf	Cancer, Scorpio, Pisces
Spring onion	Earth		Root	Taurus, Virgo, Capricorn
Swede	Earth		Root	Taurus, Virgo, Capricorn
Sweetcorn	Fire		Fruit/seed	Aries, Leo, Sagittarius
Thyme	Water		Leaf	Cancer, Scorpio, Pisces
Tomato	Fire		Fruit/seed	Aries, Leo, Sagittarius
Turnip	Earth		Root	Taurus, Virgo, Capricorn
Vine	Fire		Fruit/seed	Aries, Leo, Sagittarius

The Moon and the Weather

Any subject connected to the moon quickly becomes a magnet for myth, superstition, folklore and wild conjecture. This is as true for the weather as it is for agriculture, medicine and astronomy.

The argument that the moon must affect the weather is a familiar one. It says that since the moon's gravitational force creates tides in the sea, it must also generate similar tides in the atmosphere.

For 2000 years, before meteorological instruments, weather forecasting was a matter of experience, guesswork, superstition and myth, and the moon played its usual large part.

Weather forecasting as a science began to emerge at the end of the eighteenth century, with the development of the barometer and other instruments. Today, the most important tool in the meteorologist's box is the weather satellite. What the telescope was to astronomy, the satellite is to meteorology.

Accurate weather forecasting has only been possible since the Second World War, when an ability to predict the weather became a matter of military survival. Meteorologists played important parts in the planning and execution of enormous battles like the invasion of Europe on D-Day. Weather patterns are very fluid, and even with weather satellites making consistently accurate forecasts, predictions for more than a few days ahead are very difficult.

Accurate weather forecasting has only been possible since the Second World War, when an ability to predict the weather became a matter of military survival.

The first treatise on the weather was *Meteorologica*, written by Aristotle in about 340 BC. Aristotle conjectured on the influence of the planets on the weather. His work was added to by Theophrastus (371–287 BC), who wrote *A History of Physics*, which included sections on 'The Winds' and 'Water, Winds and Storms'. This work influenced weather forecasting for nearly 2000 years. The Greeks were the first to acknowledge that the fluidity of weather systems would always be a problem in weather prediction.

In the second century AD, Ptolemy published a weather calendar known as a *Parapegma* which listed astronomical phenomena. Ptolemy too saw the problems of weather prediction and for him it was as much a branch of astrology as anything else.

Seven hundred years later, the Arab philosopher Jaqub ibn Ismaq al-Kindi (800–870) wrote a treatise on weather forecasting that included the idea that the moon's place in the zodiac was an important factor in weather forecasting.

By the fourteenth century, Europe was in the grip of the Little Ice Age. Erratic and unexpected weather patterns caused crop failures and large-scale famine. Weather forecasting became even more important.

In 1573, the astronomer Tycho Brahe published *De Nova Stella* (*On the New Star*). Brahe gave great weight to the importance of the moon's influence on the weather. The fact that the moon is so close to the earth accounted for much of its importance. Brahe thought that the positions of the constellations should be a factor in weather forecasting. Like many before him, he realized that the fluidity of the weather system would complicate the influence of the heavenly bodies.

By the seventeenth century, weather-related instruments had appeared, most notably the barometer designed by Evangelista Torricelli (1608–47). As empirical analytic scientific methods grew in importance, the notion that astrology would have anything to do with weather forecasting began to be discredited.

In 1772, Luke Howard (1772–1864) was born. Howard made detailed weather records and became known as the Father of Meteorology. His paper to the Askesian Society in about 1802 proposed a system of cloud classification that is still in use today. Howard made extensive use of the barometer and he speculated in a work called *Barometrographica* that the rise and fall of the mercury showed a cyclical pattern that could be linked to the gravitational effect of the moon on the earth's atmosphere. He presented an analysis of records covering eighteen years and used the influence of the relative positions of the moon, sun and earth to explain weather changes.

His paper to the Askesian Society in about 1802 proposed a system of cloud classification that is still in use today.

By the mid nineteenth century there were many theories about the weather. Weather forecasters themselves were a mixed band of scientists, amateurs and chancers.

The Greek philosopher and meteorologist Theophrastus had declared that 'life is ruled by fortune, not wisdom'. Two thousand years later it was fortune that was to influence the course of weather forecasting. Stephen Martin Saxby (1804–83) was a naval architect and amateur meteorologist. He used the moon to formulate his weather predictions, though he was at pains to point out that there was nothing astrological about his theories. In 1868, Saxby predicted that an important storm would occur somewhere on the east coast of North America. He did not say where the storm would break, but he did say that it would arrive in October. On 4 and 5 October, a huge weather front developed in the North Atlantic and made landfall in the Maine/New Brunswick area. It caused chaos. Life was lost and property damaged, and rainfall records were set that have still not been exceeded. It made Saxby's reputation and emphasized the popular view of the importance of the moon as an element in weather prediction.

He used the moon to formulate his weather predictions, though he was at pains to point out that there was nothing astrological about his theories.

In the nineteenth-century world of amateur science, first-rate scientific pioneers rubbed shoulders with crackpots and eccentrics. To some the use of the moon as part of a scientific theory of the weather smacked of astrology.

In 1845, the British government set up the Meteorological Department under the direction of Captain Robert Fitzroy. Fitzroy believed that the moon and the sun combined to produce an undetectable but important pull on the atmosphere. This luni-solar pull reinforced the known meteorological forces – the circulating currents caused by hot air rising over the equator and sinking at the poles. Fitzroy's personality was fragile. He was criticized for his methods and lack of expertise. His troubles were made worse by debt, and Fitzroy eventually killed himself at his house in Upper Norwood on 30 April 1865. To the Victorians this was a sign of madness, and his tragic death damaged the cause of meteorology.

Lunarists sought to explain the moon's influence in terms of electrical, magnetic and gravitational disturbances to the atmosphere that they believed were caused by alignments of the sun, moon and earth. They also sought respectability and they found it in the unwitting figure of Sir William Herschel (1738–1822) and his son John (1792–1871).

William Herschel was the builder of a giant 40-foot reflecting telescope. He had observed the moon and measured the height of lunar mountains. He was interested in the effect of the planets on the weather. He was especially interested in weather forecasting as a means to predict clear nights when he could use his telescope. In 1835, J. C. Loudon published a compendious *Encyclopaedia of Gardening*, which contained a section about the moon and its effect on the weather. The same volume reproduced a chart entitled 'Herschel's Table of the Weather'. The table claimed to be a guide to the weather, designed by William Herschel, using the position of the moon. Although both William and John had talked privately on these matters, both were anxious to distance themselves from overt lunar speculation. The problem was made worse later in the nineteenth century when John Herschel became involved in what is known as the Great Moon Hoax. It was claimed that he had observed bat-like humanoid creatures living on the moon. The rumours were false, but nevertheless damaged his reputation.

Astro-meteorological speculation about what the moon does to the weather continues to the present day. It still hovers uneasily between science and superstition. The old idea, that because the moon can be seen to influence the tides it must also exert a force over anything that is vaguely watery, prevails. The moon probably does affect the weather, but its influence is only part of the picture.

The moon probably does affect the weather, but its influence is only part of the picture.

The problem that Aristotle and Theophrastus identified is still a problem today: the weather is very precarious. Rising currents of hot air meeting falling currents of cold air are very difficult things to pin down and they are in a state of constant flux. After 2000 years of meteorological investigation we still can't look more than three days into the future.

TRADITIONAL WEATHER LORE AND THE MOON

The difficulty of predicting the weather is demonstrated by the huge amount of folk weather lore that has evolved since the first astrologers studied the movements of the moon in Mesopotamia five millennia ago. Some traditional beliefs include:

▶ Storms end at the rising or setting of the sun or moon.
▶ A large red moon with clouds will bring rain in twelve hours.
▶ Fair weather follows a moon that rises large and bright at sunset.
▶ Mists round a waxing or waning moon indicate rain.
▶ If the moon has a halo, snow and wind are coming.
▶ A fog and a small moon soon bring wind from the east.
▶ The full moon eats the clouds.
▶ An eclipse of the moon will bring tempestuous weather.
▶ Bright moon on a winter night will bring a hard frost.
▶ Black spots on the moon mean rain.
▶ Red spots on the moon bring wind.
▶ Rain will come when the moon changes in the morning.
▶ If the moon changes on Saturday it will bring rain.
▶ If the moon is full on a Saturday it will rain.
▶ A new moon at midnight means a fine month.
▶ If the weather on the sixth day of the moon is the same as the weather on the fourth day of the moon that weather will continue for the rest of the month.
▶ When the moon changes with the wind in the east the weather that month will be bad.
▶ Moon in the north brings cold; moon in the south brings warm and dry.
▶ If the moon's crescent horns point up during the last quarter there will be rain.
▶ Two full moons in a calendar month bring on a torrent.
▶ No storms when the moon is nearly full.
▶ If the waning moon brings a cloudy morning it will bring a fine afternoon.
▶ Wind will come if the moon's horns are sharp; if they are sharp in winter there may be frost.
▶ The higher the moon, the higher the clouds, the finer the weather.

Chapter 5

Astronauts, Cosmonauts and Lunar Exploration

Getting to the Moon

THE ROCKET

'Escape velocity' is the term that describes the speed a vehicle must reach in order to leave the gravitational pull of a planet. Escape velocity is a complicated and misunderstood idea. It has a precise scientific meaning, but in popular terms it means that the speed required to leave the earth's gravitational pull is about 7 miles per second, or 25,000 miles per hour. Only a rocket is capable of going that fast. Most of the development work on the rocket has taken place in the last 100 years.

ROCKET SCIENCE BEFORE THE TWENTIETH CENTURY

Until the twentieth century, rockets were used for armaments or entertainment. They are known to have been used by the Chinese against the Mongols in 1242 and they appeared on the battlefields of Europe from the eighteenth century. There is one probably apocryphal account of their use as transport. Wan Hu, a Chinese official in the sixteenth century, is described as attempting flight on a chair powered by forty-seven rockets. The contraption exploded and nothing was ever found of Wan Hu or his chair. Had he succeeded he might have claimed to have been the first astronaut.

ROCKET SCIENCE AND LUNAR EXPLORATION 1900–45

Rocket-powered flight was first seriously considered in 1919 by an American, Professor Robert H. Goddard. Goddard wrote a pamphlet entitled *A Method of Reaching High Altitude*. Goddard proposed to detonate, on the moon's surface, a magnesium flare large enough to be seen from the earth. After a series of experiments he calculated that he would need 14.5 pounds of magnesium to create a flash that would be visible on earth, and that this payload would need a rocket with a fuelled launch weight of 15 tonnes to achieve escape velocity.

In 1926, Goddard's first missile took off. It weighed 16 pounds. The rocket reached a speed of 62 miles an hour. The flight lasted 2.5 seconds and it attained an altitude of 41 feet. Goddard's ground-breaking research was still being used by rocket scientists in the 1950s.

Goddard's booklet was read by the German scientist Herman Oberth, who in 1923 had written *The Rocket into Interplanetary Space*. Oberth's idea was to construct a rocket that would circle and photograph the far side of the moon.

In the 1920s and 1930s, other scientists were speculating on the possibility of space travel. In Italy in 1923 Luigi Gussalli wrote a book, *Is it Already Possible to Attempt a Trip from the Earth to the Moon?*, proposing the use of a multistage rocket. The British Interplanetary Society was formed in 1933. Just before the Second World War it published ideas for a 100-foot-high rocket that would send three men to the moon. They would travel in a lunar lander that was similar in shape to that used by NASA in 1969. In the Soviet Union the Rocket Research and Development Centre was founded in 1924 and in 1933 launched a liquid-fuelled rocket which reached a height of 262 feet. That rocket programme was under the supervision of Sergei Korolov, who in 1961 would be the mastermind behind the flight of the first man in space, Yuri Gagarin. Known as Chief Designer, Sergei Korolov was

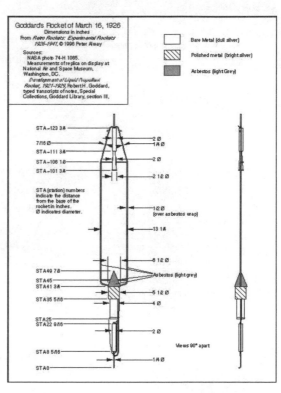

Launched on 16 March 1926, Goddard's rocket was the first to use liquid fuel.

SATURN V APOLLO FLIGHT CONFIGURATION

the driving force behind the Russian attempts to land a man on the moon.

In the Second World War much rocket research was done, especially in Germany. The German plan was to develop rockets with explosive warheads that could reach as far as New York. The work was carried out under the supervision of rocket genius Wernher von Braun. His research was very successful. By 1944, V2 rockets were bombarding London, and 60,000 slave labourers were sent to work in the rocket production plant and concentration camp at Peenemünde, where 20,000 of them died. Some of those prisoners managed to interfere with the work of von Braun and his colleagues by sabotaging the electronic guidance systems. The war ended in 1945, saving New York from the fate that befell London.

At the end of the war, von Braun surrendered to the Americans. He persuaded key members of his team to go with him. They took most of the important equipment and information with them. Many of those left behind, and a lot of the remaining equipment, were captured by the Russians. The German rocket scientists would become important elements in the race to the moon. The know-how that Hitler had thought to use to conquer the world would be used to conquer space.

1947–77

The 1950s and 1960s saw the development of rockets capable of carrying very large payloads. In the US, Wernher von Braun and his team developed the Saturn family of rockets. It would be the Saturn V which would carry the Apollo programme. In the USSR, Vladimir Chelomei designed the Proton rocket. Developed from an intercontinental ballistic missile for delivering nuclear warheads, the Proton was very successful and versions of it are still in use today. These rockets enabled the Russians and the Americans to compete in a race to land the first men on the moon. America won the space race on 20 July 1969, when Neil Armstrong became the first man to put his foot on lunar soil. Over the next three years America landed five more teams on the moon. In 1977 the Apollo programme was abandoned. All those instruments left on the moon that were still working were turned off. The space race was over.

LUNAR EXPLORATION
1978 TO THE PRESENT DAY

For about fifteen years after the last Apollo landing not much happened on or around the moon.

In 1983 the moon's gravity was used to accelerate the International Comet Explorer (ICE), which spent fifteen months orbiting the moon. ICE came to within 73 miles of the moon's surface before setting off on its 12.5-million-mile journey to examine Halley's Comet.

In 1990 interest in the moon began to reawaken. On 24 January 1990 the Japanese launched two lunar probes, Hiten and Hagoromo, and in December 1990 the Americans sent the probe Galileo to photograph the moon's surface.

In 1994 the deep-space probe Clementine spent two months in lunar orbit mapping the moon's surface. The result was the most comprehensive survey of the moon ever to be achieved in the twentieth century.

Since then the pace has quickened. America, China, Europe, India, Japan and representatives of private enterprise all plan to exploit the moon's resources. The moon is seen as a mineral resource and a jumping-off point for interplanetary exploration. Mars is our next destination and the moon is the first stop on the way. The moon's resources would be used to supply and build lunar bases. Mined material would be taken back for use on earth. There is special interest in the substance Helium-3, rare on earth but available in the lunar regolith. Helium-3 could be used as a terrestrial power source. The Chinese cosmo-chemist Ouyang Ziyuan, who is in charge of the Chinese lunar exploration programme, has said, 'Each year three space shuttle missions could bring enough fuel for all human beings on earth.'

The moon is seen as a mineral resource and a jumping-off point for interplanetary exploration.

In the twenty years since Japan's probes were launched there have been more than fifteen unmanned missions that have landed on the moon or flown round it with photographic and mapping equipment. In time bases will be established, laboratories and observatories set up and mines dug.

Mining sufficient quantities of material from the moon will require enormous amounts of explosives. The mining activity combined with human habitation will bring an atmosphere to the moon and will radically alter its surface. For as long as man has been able to look at it the moon has hovered in a near perfect vacuum, a pristine record of the stellar activity in the 5 billion years of its life. It may contain clues to what happened at the creation of the universe. Science and industry have big plans for the moon and in carrying those out they will destroy its unique environment. The moon is waiting for us, our next rainforest.

Date	US	USSR/Russia	Europe	Japan	China	India	Private enterprise
1983	ICE fly-by						
1990	Galileo photographs moon			Hiten and Hagoromo			
1992	Galileo photographs moon						
1993				Hiten impacts moon			
1994	Clementine lunar orbit for mapping						
1996	Near Earth Asteroid Rendezvous (NEAR) mission initiated						
1998	Lunar Prospector. NEAR						HGS 1 Moon Orbiter
1999	Lunar Prospector impacts Casinin						
2001	Stardust fly-by. Mars Odyssey fly-by						
2003	Mars Global survey fly-by. Mars Express		SMART 1 launched		Chang's lunar programme announced	Lunar programme announced	

The Space Race 1947–77

'The Space Race' is the term used to describe the attempts by the United States and the Soviet Union to land a man on the moon. The race was triggered on 4 October 1957, when the Soviets launched the earth-orbiting satellite Sputnik. The starting gun proper was fired by President John F. Kennedy on 25 May 1961, when he made a speech committing the US to put a man on the moon by the end of the decade.

> *First, I believe that this nation should commit itself to achieving the goal, before this decade is out, of landing a man on the moon and returning him safely to the earth. No single space project in this period will be more impressive to mankind or more important in the long-range exploration of space; and none will be so difficult or expensive to accomplish . . .*
>
> President John F. Kennedy's speech to Congress, 25 May 1961

The race became a confused test of each country's moral, political, scientific and financial strength. The race began with the development of earth-orbiting probes and progressed to the point where someone would stand on the moon and invoke its magic powers to help defend the US from communism. When that happened the space race would be won.

The estimated cost to the Americans was $25 billion and to the Russians $10 billion, or about $150 billion in modern money.

Space race: key moments		
Date	US	USSR
1946	Radio waves bounced off moon	
1957		Sputnik
1958	Satellite Explorer 1 launched Lunar probe Pioneer 1 fails Military lunar programme approved	Luna probe fails

Year		
1959		Luna 1 makes moon fly-by Luna 2 hits moon Luna 3 photographs far side of moon
1961	Pres. Kennedy announces Apollo programme to put man on moon by end 1969	Gegarin is first man in space
1962	Ranger 3 fails	Luna 4 misses moon
1964	Ranger 7 photographs moon and crashes into it	USSR announces manned lunar programme
1965	Ranger 9 photographs moon and impacts near crater Alphonsus	Zond 3 recces far side of moon
1966	Surveyor 1 lands Lunar Orbiter becomes artificial satellite	Luna 9 makes soft landing Luna 10 orbits moon
1967	Apollo 1 crew killed in launch pad fire Lunar Orbiter 4 orbits moon's poles	Cosmonaut Komarov dies flying prototype lunar lander Soyuz 1
1968	Surveyor 8 lands on moon Apollo 7 (manned) achieves earth orbit Apollo 8 (manned) makes first lunar orbit	Zond 6
1969	Apollo 11 (manned) puts first men on moon Apollo 12 (manned) lands	Luna 15 (unmanned) crashes on moon
1970	Apollo 13 (manned) mission abandoned in lunar orbit	Luna 16 (unmanned) lands and recovers soil samples Luna 17 (unmanned) lands and deploys Lunokhod 1, first lunar rover
1971	Apollo 14 (manned) lands Apollo 15 (manned) lands	
1972	Apollo 16 (manned) lands Apollo 17 (manned) lands	Luna 20 (unmanned) recovers rock sample N-1; launch fails and ends Soviet manned lunar- landing programme
1973	RAE radio astronomy observatory orbits moon Mariner 2 fly-by	Lunokhod 2 deployed (Luna 21) Luna 24 returns soil samples to earth
1977	Voyager 1 fly-by Apollo programme cancelled and all Apollo instruments on moon turned off	

The Soviet lunar exploration programme went under the umbrella title Luna. A second space exploration programme named Zond achieved some lunar flyby.

The Luna programme

The Luna programme, sometimes referred to as *Lunik* or *Lunnik*, was a series of robot missions initiated by the Russians in 1957 and which continued until 1976. The missions were designed to study the characteristics of the moon by orbital or landing vehicles carrying remote-controlled equipment.

A total of twenty-five missions were designated Luna, although there were more which failed and which were not recognized or announced as being such. The programme was a success and gave the Russians several firsts:

▶ Luna 1, although it failed to achieve lunar orbit, became the first satellite to orbit the sun.
▶ Luna 2 became the first manmade object to impact the moon.
▶ Luna 3 took the first pictures of the far side of the moon.
▶ Luna 9 made the first soft landing and sent back the first close-up panoramic shots of the moon's surface.
▶ Luna 10 became the moon's first artificial satellite.
▶ Luna 16 sent the first robotically gathered samples of moon rock back to earth.
▶ Luna 17 and Luna 21 carried the Lunokhod vehicles, which were the first remote-controlled vehicles to examine areas of the moon's surface.

* UTC stands for Coordinated Universal Time. This is a time standard based on Atomic Time and is the same as Greenwich Mean Time. The abbreviation UTC is a universally accepted compromise that overrides the way the term is phrased in a particular language. The French for example use Temps Universel Coordonné but accept the abbreviation UTC.

The programme is estimated to have cost $4.5 billion and gleaned much valuable information about the moon's chemical composition, gravity, temperature and radiation characteristics.

Luna 1 (also known as *Mechta*)
Launched 2 January 1959, 16.41 UTC*
Moon fly-by 4 January 1959

The mission plan for Luna 1 was that it should orbit the moon recording and transmitting data. Having done this it should crash into the moon's surface. Programming errors with the launcher caused it to miss the moon by 40,000 miles and enter solar orbit, where it became the first artificial planet. It was renamed *Mechta*, which is the Russian for *dreamer*.

Luna 1 was the first manmade object to achieve escape velocity from the earth. It carried communications and tracking equipment, a magnetometer, Geiger counter, a micrometeorite detector and a scintillation counter. When it was 740,200 miles into its flight it released a 2-pound cloud of sodium gas, turning it into the first artificial comet. The orange trail from the sodium could be seen from the Indian Ocean and was photographed from the Soviet observatory at Kislovodsk. At 373,000 miles into its flight and 62 hours after launch the batteries gave out and communications were lost. The satellite still orbits the sun every 443 days.

Although the mission did not achieve all its objectives, it was still very useful. Luna 1 collected data about the Van Allen belt and the earth's radiation belt. Most importantly it discovered the phenomenon of the solar winds, which emanate from the sun and cover the entire solar system.

Luna 2
Launched 12 September 1959, 06.39 UTC
Moon impact 14 September 1959, 22.02 UTC
Impact zone near Palus Putredinis

Luna 2 carried very similar equipment to Luna 1. In flight it confirmed the observations made by Luna 1 about the solar wind and the Van Allen belt. On 14 September, after 33 hours of flight, signals from the satellite stopped, indicating that it had impacted as planned west of Mare Serenitatis near the craters Aristides, Archimedes and Autolycus. The impact was tracked by the British observatory at Jodrell Bank. As well as its scientific equipment the satellite carried two spheres covered in Soviet insignia, pictures of Lenin and inscriptions bearing the date. These became part of the debris from the first object that man had ever sent to the moon. Astronomers from the Szabadsághegy observatory in Budapest claim to have seen and recorded the impact, although this is thought to be unlikely.

Launched 4 October 1959, 02.24 UTC
Fly-by 6 October 1959, 14.16 UTC, 4000 miles

Luna 3 achieved another first: it photographed the far side of the moon.

The craft was equipped with a very complex onboard imaging system, at the heart of which were two cameras connected to two lenses. The first combination, with a 200mm, f5.6 lens, would take shots of the entire moon, and the second, with a 500mm, f9.5 lens, would take closer shots of selected areas. The cameras were loaded with forty frames of radiation- and temperature-resistant isochrome film which one Soviet source claimed had been retrieved from a

The far side of the moon, photographed by the Soviet Luna 3 probe.

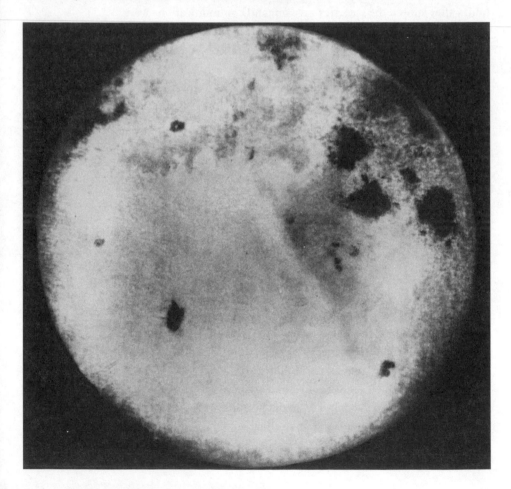

US spy balloon. The system was linked to a photocell which would detect the presence of the moon, manoeuvre the satellite so that the camera could take its shots, and then initiate the photoshoot. After the film had been exposed it would be processed and dried on board. The images were then scanned by a method known as flying spot, which converted the images on the film into electrical signals that could be transmitted to earth. There were two transmission speeds: a slow rate of 30 minutes per picture for when the earth–satellite distance was great; and a much faster rate of 15 seconds per picture when the earth–satellite distance was small. The picture resolution was 1000 × 1000 pixels.

After take-off, the satellite experienced some problems. The radio transmitted at a much lower power than had been anticipated and the

spacecraft's internal temperatures were higher than planned. This could have damaged the emulsion on the 35mm film and jeopardized the photographic mission. In spite of these problems, the satellite reached the far side of the moon on 6 October. Gas jets brought its rate of rotation to a stop. The camera port opened and the cameras began to take photographs at shutter speeds of 1/22, 1/400, 1/600 and 1/800 of a second. After making twenty-nine exposures the shutter on the 500mm camera jammed and the session was over.

The first close-up photograph of the moon, taken on 3 February 1966 by the Soviet Luna 9 just before their soft landing.

On earth, two antennae tracked the spacecraft's progress: the Kochka antenna near Simeis in the Crimea and the Petropavlovsk antenna in Kamchatka. In an attempt to improve the noise-to-signal level from the satellite, shipping in the Black Sea was ordered to maintain radio silence and the streets in Simeis round the Kochka antenna were closed to traffic. The first four attempts to transmit failed. Then, with the satellite 300,000 miles from earth, two pictures were received. As the satellite got much closer to earth fifteen more were received. The pictures were of poor quality but the Soviet scientists and technicians were able to view the far side of the moon, a sight never before seen by a human being. The images in the fuzzy black-and-white photographs showed dark spots and light patches among which were the moon's largest impact crater, the Aitken Basin. Features that could be only partly seen on the edge of the near side now revealed their whole shapes, especially the seas Humboltianum, Smythii, Australe and Marginis. One dark sea-like area was quickly named Mare Moscovrae.

Luna 4
Launched 2 April 1963, 08.04 UTC
Fly-by 6 April 1963, 13.25 UTC, 5180 miles

The precise objectives of Luna 4 are unknown, but it is thought that they included a soft lunar landing. At 13.25 UTC on 5 April the satellite achieved a lunar fly-by, missing its destination by 5800 miles. Contact was lost on 6 April.

Luna 5
Launched 9 May 1965, 07.55 UTC
Moon impact 12 May 1965, 19.10 UTC

A malfunction of the onboard gyroscope caused the satellite to spin on its axis with a subsequent loss of control. As the spacecraft approached the lunar surface, attempts to fire the braking rocket failed and it was destroyed on impact. In the two years between Luna 4 and Luna 5 there had been several launch failures.

Luna 6
Launched 8 June 1965, 07.41 UTC
Fly-by 11 June 1965, 17.00 UTC, 100,000 miles

This mission was another attempt at a soft landing. Incorrect course correction coordinates were sent to the craft, which put the spacecraft on the wrong heading, exhausting its fuel. Ground

Control slightly redeemed their position by putting the craft through a successful rehearsal of the steps to achieve a landing. On 11 June the spacecraft flew by the moon and entered an orbit round the sun. Contact was lost 3.5 million miles from earth.

Luna 7

Launched 4 October 1965, 07.55 UTC
Moon impact 7 October 1965, 22.08 UTC, W. of crater Kepler

Luna 7 was intended to soft-land on the moon. The onboard optical sensor had been incorrectly set, and the spacecraft lost altitude control on the approach to landing. The probe crashed on to the moon at 22.08 UTC on 4 October. The mission was a failure.

Luna 8

Launched 3 December 1965, 10.48 UTC
Moon impact 6 December 1965, 21.51 UTC, W. of crater Kepler

Luna 8 was yet another attempt to make an unmanned soft landing. The craft successfully entered lunar orbit. As it was being brought in to land, two cushions were inflated, one of which was pierced by a plastic mounting. The escaping gas spun the satellite. It managed to fire its retro engine for about 9 seconds and to regain some altitude. This was not enough to slow it down for landing, and it was destroyed on impact.

The mission was not a total write-off. It completed several development experiments on equipment related to communications, solar orientation, flight trajectory and onboard radio control systems.

Luna 9

Launched 31 January 1966, 11.45 UTC
Moon soft landing 3 February 1966

Inside the egg-shaped container that was the heart of Luna 9 was scientific measuring equipment, radio communications equipment and, most important of all, a new lightweight and power-efficient camera. The camera could transmit pictures at a maximum

resolution of 6000 lines. Pictures taken at this resolution then took 100 minutes to transmit to earth.

The container was carried aboard the landing delivery vehicle. The landing vehicle made a slow controlled descent on to the moon's surface, in the Oceanus Procellarum. Five metres above the lunar surface, sensors triggered a mechanism

Top: Soviet Luna 9 automatic lunar station.

Bottom: Soviet Luna 9 space probe with its stabilizing outer shell deployed.

which projected the egg several metres away from the landing ground of the main probe. The skin of the egg peeled back in four petals to stabilize the station. Four spring-loaded antennae were deployed so that the panoramic camera could begin work.

Just over four minutes after impact, the egg began to transmit data and within fifteen minutes it took its first picture of the lunar surface: a panorama of the surrounding landscape.

The pictures were of poor quality but clearly showed rocks and craters. The fact that the egg was successfully placed on the surface ended the speculation that any landing object would sink into the regolith (the layer of fine dust that covers almost the entire surface). The pictures were intercepted by the radio observatory at Jodrell Bank and with the help of equipment borrowed from the *Daily Express* were transmitted to the world, stealing the Soviet thunder.

After about eight hours the batteries gave out and the mission ended. Sadly it was not witnessed by the leader of the Soviet space programme – Sergei Korolov had died two weeks earlier. His body is interred in the wall of the Kremlin in Moscow.

Luna 10

Launched 31 March 1966, 10.48 UTC
Moon orbit 3 April 1966, 18.44 UTC

On 3 April, Luna 10 entered lunar orbit and became the moon's first artificial satellite. It carried a magnetometer, micrometeoroid detection equipment, and infra-red and radiation sensors. It also carried equipment to transmit a version of the Soviet anthem 'The Internationale to the 23rd Congress of the Communist Party of the Soviet Union. An equipment failure on the day of transmission brought disaster, and the Congress had to make do with a prerecorded version made during a rehearsal the previous day.

The probe's most important discovery was that extreme irregularities exist in the gravitational field of the moon. These are now known as mass concentrations or mascons. After completing 460 orbits in 56 days the probe's batteries gave out and the mission ended.

Luna 11

Launched 24 August 1966, 08.03 UTC
Moon orbit 27 August 1966

Luna 11 was a lunar orbiter designed to photograph the moon's surface, and to investigate the gravitational anomalies and the radiation environment of deep space. Immediately after the probe entered lunar orbit, debris blocked a guidance jet nozzle. The probe could not manoeuvre to point the onboard camera at the moon; instead it pointed at the blackness of deep space. The probe stayed in orbit for five weeks until contact was lost on 1 October 1966. During that time it investigated the moon's magnetic field, used gamma and X-ray spectrometers to analyse elements of the moon's surface and collected data on micrometeoroids and the ability of the moon's surface to reflect radio waves.

The probe stayed in orbit for five weeks until contact was lost on 1 October 1966.

It achieved 277 orbits and made 137 radio transmissions. It also carried out a mechanical test on a gear reduction system that was destined to be used on the planned unmanned Lunokhod lunar rover. It found that, in spite of the high loads to which it would be subjected, the system functioned very adequately in a vacuum.

Luna 12
Launched 22 October 1966, 08.38 UTC
Moon orbit 26 October 1966, 20.45 UTC

Luna 12 completed the photographic mission attempted by Luna 11. The images were to be used in support of a manned landing. As with earlier missions, the film would be processed and scanned on-board and transmitted to earth. Wary of the way Jodrell Bank had intercepted the pictures from Luna 9, the Russians transmitted them on two different frequencies. They then began to jump between the two frequencies in a predetermined and secret sequence. Jodrell Bank needed a day to be reprogrammed to new frequencies. No pictures were intercepted and the Russians released only two to the rest of the world.

With the photographic-mapping element of the voyage complete, Luna 12 used its onboard scientific equipment to examine the lunar surface and solar wind. It was judged that the regolith had a density of 1400 kilos per cubic metre.

Spare fuel was used to modify the probe's orbit. It made another 602 orbits before contact was lost on 16 January 1967, and the mission ended.

Luna 13

Launched 21 December 1966, 10.17 UTC
Moon soft landing 24 December 1966, 18.04 UTC

On 24 December 1966 Luna 13 landed in the region of Oceanus Procellarum, one of the smoothest parts of the moon. Like Luna 9, the landing stage carried an egg-shaped container which could be projected away from the main craft to rest on the surface, where it would stabilize and deploy its equipment. The main aim of the mission was to examine the regolith. Luna 13 carried a Gruntometer penetrometer and a Plotnomer densitometer. Both were deployed on jointed arms about 5 feet in length. The Gruntometer used an explosive charge to dig about 2 inches into the regolith. Other equipment measured the surface temperature and analysed alpha rays scattered about by the densitometer. Two Volga cameras were carried, and although one of these failed, the other managed to take five panoramic views of its surroundings. Contact with the probe was lost on 28 December.

Luna 14

Launched 7 April 1968, 10.09 UTC
Moon orbit 10 April 1968, 19.25 UTC

Little is known about this mission. Its announced objectives were to study lunar gravitational anomalies, measure solar winds and cosmic rays, and carry out radio communications experiments. No camera was carried on the mission. Some of the data from the mission would have been used to establish future orbits for the unmanned and manned missions that were planned in the near future.

Luna 15

Launched 13 July 1969, 02.54 UTC
Moon impact 21 July 1969, 15.47 UTC

This unmanned mission was an attempt to upstage the US manned landing of Apollo 11. The mission objective was to retrieve soil from the moon's surface and return it to earth. Launch took place three days before Apollo 11 and there was concern among the Americans that the two missions would jeopardize each other. The Soviets said this could not happen as they were communicating on different frequencies. In the early morning of 21 July 1969, Neil Armstrong became the first man to step on to the moon and into history. Eleven hours later Luna 15, travelling at 300 miles per hour, smashed on to the Sea of Crises and into oblivion.

Luna 16
Launched 12 September 1970, 13.25 UTC
Moon soft landing 20 September 1970, 05.18 UTC

This was the first Soviet probe to land on the moon and return samples to earth. The spacecraft was designed in two stages, a descent and an ascent module combined. Once landed, the ascent module would use the descent module as a launch pad. The descent stage was equipped with TV cameras, radiation- and temperature-measuring devices, and communications equipment. It entered lunar orbit on 17 September 1970 and spent a day in orbit studying the moon's gravity. At 15.18 UTC it landed to the north-east of the Sea of Fertility. Drilling commenced and achieved a depth of 35 centimetres. The sample was placed in a spherical container mounted on the ascent stage; some material was lost in the storage phase. After 26 hours and 25 minutes on the lunar surface the ascent stage took off and headed back to earth carrying 3.7 ounces of soil. It left behind the descent stage, which continued to transmit data about the temperature radiation phenomena on the surface. Landfall was successfully made outside the town of Dzhezkazgan in Kazakhstan. The Soviets had achieved the first robotic recovery of a soil sample from any extraterrestrial body.

Luna 17
Launched 10 November 1970, 14.44 UTC
Moon soft landing 17 November 1970, 03.46 UTC,
Sea of Rains

Luna 17 carried the first moon rover to be deployed on the moon. It was called Lunokhod 1. *Lunokhod* is Russian for 'moonwalker'.

The development of the Lunokhod

In 1954, the first Russian proposals for an unmanned lunar laboratory appeared. In 1960 work began on a vehicle with a weight of about 1400 pounds, which would be transported by an N Series rocket with a launch lift capability of 16 tonnes. The chassis was to have been built under the supervision of the State Committee of Agricultural and Tractor Machine Building but the contract went in the end to the Mobile Vehicle Engineering Institute VN11, or 'Trans Mash', who were authorized to start work on 3 August 1964.

Several prototypes were tried, starting with a four-wheel vehicle coded SH-1 (the SH stood for *shassi*, which is Russian for chassis). In the next six years the rover was modified and the launch vehicle changed to the Proton heavy-lift rocket. In 1967 an eight-wheel prototype emerged. Known as the Lunokhod 1, the vehicle

»

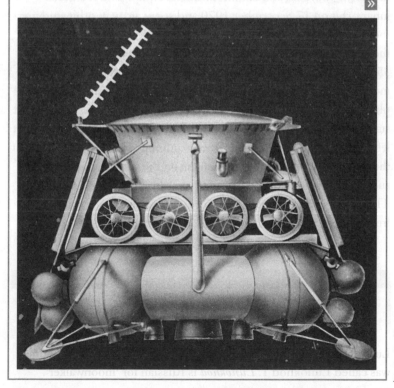

The Luna 17 space rover with its payload Lunokhod 1 ready for deployment.

weighed 1666 pounds, and was 53 inches high and 84.5 inches in diameter. Lunokhod 1 was to be controlled by a specially trained crew of five based at the Deep Space Communications Centre outside Moscow. The operating crew consisted of a commander, a driver, a systems engineer and a radio operator. Once on the moon the landing vehicle deployed two ramps down which the rover would crawl on to the surface.

The vehicle was powered by solar batteries which could be recharged during the day. Electric motors delivered power to the eight wheels. Should they fail, the motors could be disconnected by small explosive charges allowing the wheels to freewheel. A ninth wheel was attached to the rear of the vehicle to act as an odometer measuring the distance travelled. The wheels were of light metal mesh built by the Kharkov State Bicycle Plant. The Lunokhod had sensors which could stop it if its angle of tilt became too severe. It had a top speed of 1.24 miles per hour.

Lunokhod's equipment was carried in its body, protected by a hinged lid that had solar cells on the underside. The lid would be opened during the day to allow regeneration of the batteries and closed at night to conserve heat. The equipment chamber was heated by the radioactive decay of a small quantity of Polonium-210. The vehicle carried a computer, sets of cameras to take panoramic and directional views, X-ray and cosmic ray telescopes, and equipment for assessing the composition of the lunar surface. Communications with earth would be by a laser telemetry unit designed and built in France by Sud Aviation Cannes.

On 19 February 1969 the Russians made their first attempt to launch a Lunokhod, but the launch vehicle blew up and the Lunokhod was destroyed. This was kept a secret for many years.

Launch of second Lunokhod 1

On 10 November 1970 Luna 17 took off carrying a replacement Lunokhod 1. On 17 November the mission landed near the northern rim of Mare Imbrium (Sea of Rains), close to the Gulf of Rainbows. Two hours later the ramps were deployed and Lunokhod 1 crawled out on to the lunar surface.

Lunokhod had been designed to operate for 90 earth days. But in fact it did much better, and continued operating for eleven months. During its time on the moon's surface it took 20,000 guidance photographs and 206 panoramas; mapped 96,000 square yards of the lunar surface; and made 500 soil measurements and twenty-five measurements of the lunar regolith. The machine had mechanical problems which were the equivalent of having its handbrake jammed on. Even so, it managed to travel 10.54 kilometres.

On 4 October 1971, the fourteenth anniversary of the launch of Sputnik, the Russians closed the Lunokhod project. It had been a very successful mission.

Luna 18

Launched 2 September 1971, 13.40 UTC
Moon impact 11 September 1971, 07.48 UTC

This probe made fifty-four lunar orbits before it landed on 11 September 1971. After it set down in very difficult mountainous terrain, all contact with it was lost.

Luna 19

Launched 28 September 1971, 10.00 UTC
Moon orbit 3 October 1971 (exact time not known)

Luna 19 was an unmanned orbiter, with the task of capturing panoramic views of the moon's mountainous areas. It continued the programme undertaken in earlier lunar missions of identifying, mapping and investigating the magnetic anomalies known as mascons. It entered lunar orbit on 3 October 1971 and before it was shut down in October 1972 it had completed over 4000 orbits of the moon.

Luna 20

Launched 14 February 1972, 03.27 UTC
Moon landing 21 February 1972, 19.19 UTC

Luna 20 landed in the Apollonius Highlands near the Mare Fecunditatis. Before it began its main mission of sampling, the onboard camera was used to photograph the sky to identify the earth's position and to record the ground which was covered in regolith-free rocks. The ground proved to be very hard. The initial bore was 4 inches deep and sample drilling had to be periodically halted to allow the drill bit to cool. After 40 minutes the drilling was abandoned completely. The drilling equipment had achieved a depth of only 6 inches and had taken a sample of only one ounce.

The ascent stage took off on 23 February. It had been on the moon's surface for less than 28 hours. On 28 February the capsule descended through a snow storm towards the semi-frozen Karakingir river, 40 miles north of Dzhezkazgan in Kazakhstan. It landed on a small island and Soviet recovery helicopters quickly spotted its high-visibility parachutes and recovered the capsule.

The tiny amount of rock it carried was invaluable. It came from the Highlands and was composed of high volumes of aluminium and calcium oxides, anorthositic rock that had once been part of the lunar magma. Analysis showed that these samples contained rock which was 3 billion years old.

Luna 21

Launched 8 January 1973, 06.55 UTC
Moon soft landing 15 January 1973, 22.35 UTC

Luna 21 took off carrying Lunokhod 2, a modified version of Lunokhod 1. The journey to the moon was beset with problems. Communications were bad and Lunokhod's hood had to be opened in flight to allow its solar cells to recharge the lunar lander.

On 15 January at 22.35 Lunokhod landed in Le Monnier crater. Its task was to photograph the lunar surface; to determine the feasibility of telescopic observations from the moon; to study the properties of the lunar surface; and to measure solar X-rays and magnetic fields.

Lunokhod carried three TV cameras, with one mounted at eye level for navigation purposes. These cameras had a multispeed recording capability of 3.2, 5.7, 10.9 or 21.1 seconds per frame. This is much slower than a conventional TV camera, which records at 25 frames per second (0.04 seconds per frame), but much faster than Lunokhod 1, which had a fixed rate of 21.1 seconds per frame. The vehicle had the same heating and power arrangements as Lunokhod 1 and was equipped with more instruments. Extra cameras had been added, an ultra-violet telescope, a magnometer, and an improved retro-reflector, again supplied by the French. The equipment for measuring the lunar surface had also been modified. This was called RIFMA: Roentgen Isotopic Fluorescence Method of Analysis.

Luna 24 about to land and collect samples of lunar soil.

Lunokhod 2 deployed soon after landing and immediately suffered problems with its navigation system. These could have threatened the mission, but NASA had unofficially supplied the Russians with high-resolution pictures taken by Apollo 17. The maps helped solve the navigational problems.

Lunokhod 2 functioned well. It travelled by day using the sun to recharge its batteries and it hibernated at night. It took nearly 80,000 guidance pictures and eighty-six panoramas, and discovered that during the lunar day the ambient light was made brighter by lunar dust suspended in the atmosphere. In the four months of the mission it travelled 23 miles through hilly terrain and at one point was up to its axles in regolith. It set a lunar record, covering 1255 yards in 6 hours. Lunokhod performed many tests on the lunar soil. After four months it malfunctioned and its journey ended. The retro-reflectors mounted on its chassis are still operational today. Lunokhod's current position on the moon is known to the exact yard.

Luna 22

Launched 29 May 1974, 08.57 UTC
Moon orbit 2 June 1974 (exact time not known)

Luna 22 was one of the longest and most successful missions of the Soviet lunar programme. Like Luna 19 it studied anomalies in the moon's magnetic field, to measure the solar winds. It was still transmitting pictures after fifteen months in orbit. The mission was terminated in November 1975.

Luna 23

Launched 28 October 1974, 14.30 UTC
Moon soft landing 6 November 1974 (exact time not known)

An unmanned lander, the mission's objective was to retrieve soil samples and return with them to earth. The sampling equipment was damaged on landing and no samples were returned. The lander transmitted for three days, during which time the Soviets cobbled together a scratch programme of scientific investigation.

Luna 24

Launched 9 August 1976, 15.04 UTC
Moon soft landing 18 August 1976, 06.36 UTC

Luna 24 landed in the Sea of Crises (Mare Crisium) on 18 August 1976. The Sea of Crises is the site of a large lunar mascon. Luna 24 was equipped with a modified drill system designed to dig deeper than Luna 16 or 20 and to retain its sample, which would be fed into a flexible plastic tube. The drill penetrated just over 2 yards, which is what it had been designed to do. The ascent stage took off the same day and landed back on earth on 23 August. It landed at a new recovery area: Surgut in Siberia. The samples recovered were shared with US and British scientists. They contained small pebbles of iron-rich material about a third of an inch in diameter.

Lunokhod 3

This vehicle carried movable stereoscopic cameras and other modifications. It never left the earth and can be seen today in the NPO Lavochkin Museum in Moscow.

Luna missions

Probe	Launch date	Mission	Outcome
Luna 1	02.01.1959	Lunar impact	Missed moon and entered solar orbit
Luna 2	12.09.1959	Lunar impact	Achieved lunar impact
Luna 3	04.10.1959	Lunar fly-by	Took first pictures of far side of moon
Luna 4	02.04.1963	Lunar landing	Missed moon by 5000 miles
Luna 5	09.05.1965	Lunar landing	Crashed on landing
Luna 6	08.06.1965	Lunar landing	Missed moon by 100,000 miles
Luna 7	04.10.1965	Lunar landing	Crashed on landing
Luna 8	03.12.1965	Lunar landing	Crashed on landing
Luna 9	31.01.1966	Lunar landing	Landed and took pictures
Luna 10	31.03.1966	Lunar orbit	Achieved orbital scientific exploration
Luna 11	24.08.1966	Lunar orbit	Achieved orbital scientific exploration

Probe	Launch date	Mission	Outcome
Luna 12	22.10.1966	Lunar orbit	Achieved orbital scientific exploration
Luna 13	21.12.1966	Lunar landing	Landed and took pictures
Luna 14	07.04.1968	Lunar orbit	Achieved orbital scientific exploration
Luna 15	13.07.1969	Return samples to earth	Crashed on landing
Luna 16	12.09.1970	Return samples to earth	Returned 3.7 ounces of moon dust to earth
Luna 17	10.11.1970	Land and deploy Lunokhod 1 lunar rover	Deployed rover, which travelled 6.5 miles
Luna 18	02.09.1971	Return samples to earth	Crashed on landing
Luna 19	28.09.1971	Lunar orbit	Achieved orbital scientific exploration
Luna 20	14.02.1972	Return samples to earth	Returned 1.8 ounces of moon rock to earth
Luna 21	08.01.1973	Land and deploy Lunokhod 2 lunar rover	Deployed rover, which travelled 23 miles
Luna 22	29.05.1974	Lunar orbit	Achieved orbital scientific exploration
Luna 23	28.10. 1974	Return samples to earth	Landed but malfunctioned. No samples returned
Luna 24	09.08.1976	Return samples to earth	Returned 6 ounces of moon dust to earth

THE US MOON PROGRAMME

The US moon programme started very badly and suffered many setbacks. It often looked as though they would be left behind and that the Soviets would win the space race. Between 1958 and 1964, fifteen consecutive unmanned lunar flights failed to achieve their mission objectives. But in the end America prevailed.

The moon programme developed in five stages, with each stage attempting to build on the achievements of the last. The stages were:

- The Pioneer programme. Designed to achieve lunar orbit or fly by the moon. All the missions between 1958 and 1960 failed.
- The Ranger programme. A series of nine probes, building towards lunar impact. Although this programme had a very high failure rate it saw the beginnings of success in US moon exploration.
- The Surveyor programme. Nine missions to return photographs and geological data from the moon's surface. Very successful.
- The Lunar Orbiter programme. Five probes sent to make extensive photographic coverage of the moon.
- The Apollo programme. Seventeen manned and unmanned missions to explore and return samples from the surface of the moon.

Pioneer programme
1958–60

The nine Pioneer missions in this period failed. The programme was beset by problems at launch and in flight.

Ranger programme
1961–5

There were nine Ranger missions. Each would impact on the moon and photograph the surface in the last thirty minutes or so of flight. In spite of early disasters and a very high failure rate the programme provided over 15,000 photographs of the moon's surface, at a cost of around $170 million.

These were the first spacecraft to utilize 3 Axis Attitude Stabilization. This meant the probes did not spin and could retain a fixed relationship with the moon, sun, earth and stars.

A Ranger lunar probe prior to impact.

Most of the probes carried the same equipment: six cameras, deployed in two channels – P (partial) and F (full). Each channel operated independently of the other. The P channel had two cameras, a wide-angle and a narrow-angle. The F channel had four cameras, two wide-angle and two narrow-angle. Between them the cameras would record everything except the last 0.22 seconds of impact.

With Ranger, NASA learnt a lot about rocket delivery systems but missed many scientific objectives.

Ranger programme			
	Launch date	**Mission**	**Outcome**
Ranger 1	23.08.1961	Lunar impact	Failed to make orbit. Good test of equipment
Ranger 2	18.12.1961	Lunar impact	Failed to achieve orbit
Ranger 3	26.01.1962	Lunar impact	Missed moon
Ranger 4	23.04.1962	Lunar impact	Impact. Failed to send pictures
Ranger 5	18.10.1962	Lunar impact	Missed moon
Ranger 6	30.01.1964	Lunar impact	Impact. Failed to send pictures
Ranger 7	28.07.1964	Lunar impact	Impacted, sending 4308 pictures
Ranger 8	17.02.1965	Lunar impact	Impacted, sending 7137 pictures
Ranger 9	21.03.1965	Lunar impact	Impacted, sending 4308 pictures

The Surveyor programme
1966–8
The Surveyor programme's objectives were to prove that lunar landings were possible by soft-landing seven probes on the moon's surface. The probes were unmanned and due to land at different locations. Once down, the Surveyors would photograph and investigate the nature of the lunar surface. Five of the seven made soft landings as planned. Surveyors 2 and 4 crashed on the moon's surface.

Surveyor 1
Launched 30 May 1966
This probe landed on 7 June. It was not specifically intended to carry out scientific experiments. Nevertheless, it carried two television cameras and over 100 sensors. It landed according to plan in the Oceanus Procellarum and succeeded in sending back 11,000 pictures and many data about the surrounding surface. The mission ended on 14 July 1966. Engineering contact was maintained until 7 January 1967.

Surveyor 2
Launched 20 September 1966

Control was lost during a mid-course correction. The craft lost stability and crashed into the moon on 23 September 1966.

Surveyor 3
Launched 17 April 1967

Surveyor 3 landed on 20 April. On landing, it bounced three times, reaching a height of 35 feet before safely settling in the Mare Cognitum area of the Oceanus Procellarum. In spite of the rigours of landing, it performed almost perfectly. It sampled the top 7 inches of the regolith and sent back 6315 images. It was suspected that the probe's camera had been infected on earth by the bacterium *Streptococcus mitis*. The bacterium survived the moon's environment until it was recovered by Apollo 12. This led to more rigorous sterilization methods on future missions.

Surveyor 4
Launched 14 July 1967

Surveyor 4 performed perfectly until two minutes before touchdown, when it is thought it blew up.

Surveyor 5
Launched 8 September 1967

Surveyor 5 landed safely in the Mare Tranquilitatis. Its mission lasted until 17 December 1967, during which time it transmitted nearly 20,000 images and hours of data about the lunar surface. It carried an Alpha scattering surface analyser to investigate the quantity of major elements in the lunar surface.

Surveyor 6
Launched 7 November 1967

Surveyor 6 landed in the Sinus Medii and accomplished all its objectives. It sent a record 30,027 images to earth and achieved a short 2.8-yard relocation. The probe was shut down on 24 November 1967 for the lunar night. It was reactivated on 14 December but it returned no further meaningful information.

Surveyor 7
Launched 7 January 1968

This was the last probe of the Surveyor programme. Its mission objectives were the same as for the previous six missions. The probe was fitted with some extra equipment, including a camera with a polarizing filter, extra magnets and auxiliary mirrors which enabled it to take stereoscopic images of the area below the craft. After landing, the Alpha surface scattering equipment jammed. This was remedied by Ground Control using the soil-sampling scoop to push it into place. In spite of intermittent battery malfunction Surveyor 7 sent back 21,091 pictures, until contact was lost on 20 February 1968.

All the Surveyor probes or their remains are still on the moon.

The Surveyor programme			
	Launch date	**Mission**	**Outcome**
Surveyor 1	30.05.1966	Moon landing	First US moon landing. 11,000 pictures taken
Surveyor 2	20.09.1966	Moon landing	Crashed on impact
Surveyor 3	17.04.1967	Moon landing	Landed. Completed photographic and geological missions. 6,000 pictures returned
Surveyor 4	14.07.1967	Moon landing	Contact lost before landing
Surveyor 5	08.09.1967	Moon landing	Landed. Completed photographic and geological missions. 19,000 pictures returned
Surveyor 6	07.11.1967	Moon landing	Landed. Completed photographic and geological missions. 30,000 pictures returned
Surveyor 7	07.01.1968	Moon landing	Landed. Completed photographic, laser and geological missions. 21,000 pictures returned

Lunar Orbiter programme

1966–7

The objective of the Lunar Orbiters was to map the moon and identify potential landing sites. The five orbiters covered 99 per cent of the moon's surface. They carried sophisticated photographic equipment, including a processing and scanning capability that could transmit the images back to earth. Two lenses, a 610mm high-resolution narrow-angle and an 80mm medium-resolution wide-angle, were mounted to take matched shots of the same area. The 70mm film could be tracked to compensate for the movement of the spacecraft.

The Lunar Orbiter programme

	Launch date	Mission	Outcome
Lunar Orbiter 1	10.08.1966	Achieve lunar orbit and take pictures	80 days in orbit. Pictures taken successfully
Lunar Orbiter 2	06.11.1966	Achieve lunar orbit and take pictures	339 days in orbit. Pictures taken successfully
Lunar Orbiter 3	05.02.1967	Achieve lunar orbit and take pictures	246 days in orbit. Pictures taken successfully
Lunar Orbiter 4	04.05.1967	Achieve lunar orbit and take pictures	180 days in orbit. Pictures taken successfully
Lunar Orbiter 5	01.08.1967	Achieve lunar orbit and take pictures	183 days in orbit. Pictures taken successfully

The Apollo programme

1961–72

The Apollo missions were designed to put a man on the moon and to conduct scientific experiments to investigate the moon's geology and history. They were also to bring lunar rock samples back to earth. The Apollo missions used Saturn V rockets as the launch vehicles. The three-man crews were carried in a spacecraft that consisted of a Command Module (CM) and a Lunar Module (LM). Once in lunar orbit, the two modules separated. The Lunar Module would land and the Command Module would continue in orbit and wait to be reunited with the LM. Of the seventeen Apollo missions, six landed on the moon and most were successful in contributing to the objectives. Apollo 1 was cancelled after a launch pad fire killed the crew and Apollo 13 was aborted after an onboard explosion threatened the lives of the three crewmen.

Astronauts	Dates: launch/ moon landing/ earth splashdown	Outcome
Apollo 1 Gus Grissom Ed White Roger Chaffee	Planned launch 27.2.1967	Mission aborted after fire killed crew on launch pad
Apollo 7 Walter M. Schirra, Jr Donn F. Eisele Walter Cunningham	Launch 11.10.1968 Splashdown 22.10.1968	First manned mission. Achieved 163 earth orbits
Apollo 8 Frank Borman James A. Lovell, Jr William A. Anders	Launch 21.12.1968 Splashdown 27.12.1968	First manned use of Saturn V. Orbited moon on Christmas Eve
Apollo 9 James A. McDivitt David R. Scott Russell R. Schweickart	Launch 03.03.1969 Splashdown 13.03.1969	In-flight test of Lunar Module. Achieved 151 earth orbits
Apollo 10 Thomas P. Stafford John W. Young Eugene A. Cernan	Launch 18.05.1969 Splashdown 26.05.1969	Orbited moon and flew Lunar Module to within 9 miles of lunar surface
Apollo 11 Neil A. Armstrong Michael Collins Edward E. 'Buzz' Aldrin	Launch 16.07.1969 Moon landing 20.07.1969 Splashdown 24.07.1969	First manned lunar landing
Apollo 12 Charles 'Pete' Conrad Richard F. Gordon, Jr Alan L. Bean	Launch 14.11.1969 Moon landing 19.11.1969 Splashdown 24.11.1969	Landed on moon. Retrieved parts of Surveyor 3
Apollo 13 James A. Lovell, Jr John L. Swigert, Jr Fred W. Haise, Jr	Launch 11.04.1970 Splashdown 17.04.1970	Mission aborted in space after failure of fuel tank system
Apollo 14 Alan B. Shepard, Jr Stuart A. Roosa Edgar D. Mitchell	Launch 31.01.1971 Moon landing 05.02.1971 Splashdown 09.02.1971	Landed on moon. Achieved scientific exploration and got lost
Apollo 15 David R. Scott Alfred M. Worden James B. Irwin	Launch 26.07.1971 Moon landing 30.07.1971 Splashdown 07.08.1971	Landed on moon and deployed lunar rover
Apollo 16 John W. Young Thomas K. Mattingly II Charles M. Duke, Jr	Launch 16.04.1972 Moon landing 21.04.1972 Splashdown 27.04.1972	Landed on moon and deployed lunar rover
Apollo 17 Eugene A. Cernan Ronald E. Evans Harrison H. Schmitt	Launch 07.12.1972 Moon landing 11.12.1972 Splashdown 19.12.1972	Landed on moon and deployed lunar rover. Achieved very successful geological survey

Apollo 1

Mission no. | AS-204 (Apollo 1)

Commander | Gus Grissom

Command Module pilot | Ed White

Lunar Module pilot | Roger Chaffee

Planned launch date | 27 February 1967

Mission objectives | First manned flight of a Command Module into earth orbit. Mission cancelled after the crew was killed in an onboard fire. The fire ignited during a pre-launch training and equipment assessment test.

This mission was originally designated Apollo Saturn 204 or AS-204. It was to be the manned flight of a Command and Service Module (CSM) to test 'launch operations, ground tracking and control facilities and the performance of the Apollo Saturn Launch assembly'. The mission was scheduled to run for 14 days.

The Command Module had been delivered to NASA with a lot of unresolved faults. It was the most complex design used so far and there were serious concerns about the use of pure oxygen in the pressurized cabin.

The 27 January launch simulation was to establish that the spacecraft could operate on internal power while detached from the umbilical cords and cables of the launch gantry. NASA considered this operation to be non-hazardous.

At 1pm local time, Grissom, White and Chaffee entered the module and, after some concerns about a foul 'buttermilk smell', the craft was sealed and the conversion of the interior to a pure oxygen environment initiated. As the simulation countdown progressed, problems were encountered with the communications system, causing Grissom to exclaim, 'How can we get to the moon if we can't talk between three buildings.' Other problems developed and the countdown was held at t–20 to sort these out. At 18.30.54 an electrical fault was recorded and 13 seconds later Grissom shouted:

'Fire!'

Followed by Chaffee:

'We've got a fire in the cockpit.'

And then White:

'Fire in the cockpit.'

After more confusion Chaffee shouted:

'We're on fire, let's get out. We're burning up. We're on fire, get us out of here.'

Seventeen seconds later cabin pressure reached 29 pounds per square inch and the module ruptured. It took five long minutes to open the hatch. Inside, everything was black and confused. The astronauts were dead, killed by a combination of burns and asphyxiation.

The cause of the fire was never fully established. The module contained many fire hazards and had been shoddily assembled. Miles of vulnerable wiring, 70 pounds of non-metallic inflammable material, and 34 square feet of Velcro which proved to be explosive in a high-pressure oxygen environment all contributed to the dangers of the interior. The astronauts' suits contained flammable plastic and were able to generate high static charges. Redesigning the module delayed the Apollo programme by twenty months. After pressure from other astronauts, Mission AS-204 was renamed Apollo 1.

'If we die, we want people to accept it. We're in a risky business. And we hope that if anything happens to us it will not delay the programme. The conquest of space is worth the risk of life.'

Gus Grissom, Commander, Apollo 1

Apollo 2 (AS-203)

A flight to test the effects of weightlessness on the fuel system.

Apollo 3 (AS-202)

A flight to test the rocket and heat shield.

Apollo 4, 5 and 6

The next three Apollo missions were unmanned. Designated Apollo 4, Apollo 5 and Apollo 6, they were to test the Saturn rocket and the Command and Service Module which would carry three-man crews to and from the moon. The first two missions were very successful. Apollo 6 was a disappointment.

It attracted little attention. Five days before launch, President Lyndon Johnson announced that he would not stand for re-election. On the day of the launch, Martin Luther King was assassinated in Memphis, Tennessee.

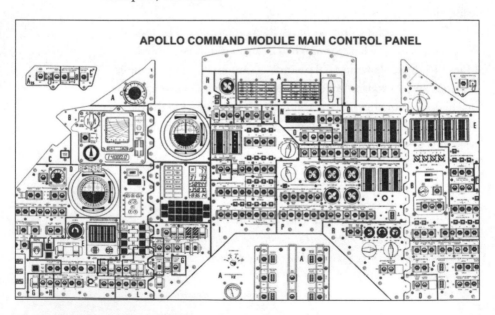

APOLLO COMMAND MODULE MAIN CONTROL PANEL

From launch the rocket malfunctioned. As it lifted off it developed a severe oscillation, which manifested itself as a stretching and shortening of the structure. This was described as pogoing. The second stage lost thrust and the third stage failed to restart. It failed to achieve its prime objective, which was to place its payload in lunar orbit.

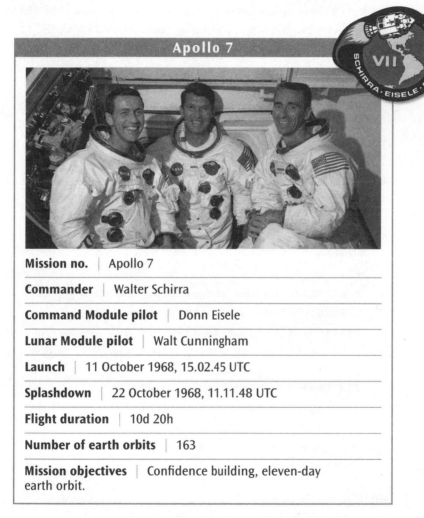

Apollo 7

Mission no. | Apollo 7

Commander | Walter Schirra

Command Module pilot | Donn Eisele

Lunar Module pilot | Walt Cunningham

Launch | 11 October 1968, 15.02.45 UTC

Splashdown | 22 October 1968, 11.11.48 UTC

Flight duration | 10d 20h

Number of earth orbits | 163

Mission objectives | Confidence building, eleven-day earth orbit.

This was the first manned flight of the Apollo programme. Its objectives were to test the Service Propulsion unit (SPU) which would put the module in and out of lunar orbit, to test the Lunar Module docking manoeuvre, and to transmit live television pictures.

The crew of Apollo 7 were the same men who had been the back-up crew for Apollo 1. Apollo 7's flight was not a total success. Conditions in the capsule were uncomfortable and over the ten-day flight relations with the ground staff deteriorated. The crew complained that they were being overworked and that they were developing colds. At one point the Capsule Communicator (the only person allowed to talk directly with the spacecraft and known as CAPCOM) asked Schirra to turn on a TV camera in the capsule.

> **SCHIRRA:** You've added two burns to this flight schedule, and you've added a urine water dump; and we have a new vehicle up here, and I can tell you at this point TV will be delayed without any further discussion until after the rendezvous.
> **CAPCOM:** Roger. Copy.
> **SCHIRRA:** Roger.
> **CAPCOM:** Apollo 7, this is CAPCOM number 1.
> **SCHIRRA:** Roger.
> **CAPCOM:** All we've agreed to do on this is flip it.
> **SCHIRRA:** With two commanders, Apollo 7.
> **CAPCOM:** All we have agreed to on this particular pass is to flip the switch on. No other activity is associated with TV; I think we are still obligated to do that.
> **SCHIRRA:** We do not have the equipment out; we have not had an opportunity to follow setting; we have not eaten at this point. At this point, I have a cold. I refuse to foul up our time lines this way.

Although the mission was declared a success, no member of this crew was ever asked to fly in space again.

BORMAN LOVELL ANDERS

Mission no.	Apollo 8
Commander	Frank Borman
Command Module pilot	Jim Lovell
Lunar Module pilot	Bill Anders
Launch	21 December 1968, 12.51.00 UTC
Splashdown	27 December 1968, 15.51.42 UTC
Flight duration	6d 03h
Lunar orbits	10

Mission objectives | To achieve lunar orbit in a manned spacecraft. Test launch capabilities of Saturn V rocket. Test communications. Recce and photograph future Apollo landing sites.

This was the first time the Saturn V launch was used to carry human beings to the moon. The noise and vibration that the crew encountered were colossal. For a time it was almost impossible to hear communications from the ground. The five F1 engines which powered the rocket burnt 20 tonnes of fuel a second. G-forces on the astronauts soared to 4.5 times those on earth and fell to zero as the first stage, the S-1C, fell away. Then the crew were slammed back into their seats as the second stage, the S-11, fired. Eight minutes later all was silent. They were in earth orbit. Over the communications system came the voice of CAPCOM Michael Collins:

'Apollo 8, you are go for TLI.'

This was the clearance that they could undertake 'trans-lunar injections', the course that would take them to the moon.

After nearly another three hours in earth orbit, the Apollo computer issued the command that would initiate the firing sequence to accelerate the rocket to escape velocity. The rockets thundered and five and a half minutes later they were on their way, travelling at 24,200 miles per hour – faster than any living thing had ever travelled.

On the way to the moon, Borman and Anders suffered from space sickness and for a while it was thought that Borman had a viral infection. If this were true, the mission would have to be aborted. After talking to Flight Surgeon Charles Berry, the mission was cleared to proceed.

The lunar orbit planned for Apollo 8 was the type known as a 'free return trajectory', which meant that in the case of problems the moon's gravity would be used to send the spacecraft back to earth. On Christmas Eve, nearly 70 hours after launch, the astronauts swung round the back of the moon. Only static came from the radios. The moon had cut them off from all contact with earth. The service module engines were fired, slowing down the spacecraft and bringing it into lunar orbit. The crew could now only wait. Below them the moon looked like wet sand. They could see the crater Tsiolkovsky, discovered nine years earlier by Luna 3. Then as their craft emerged from the moon's shadow, they were presented with another sight that had never been seen: the earth rising above the horizon of the moon. Earthrise. A sight which made Jim Lovell say:

'The vast loneliness is awe-inspiring. It makes you realize just what you have back there on earth.'

After ten orbits, the spacecraft's engines were fired and the module began its flight home. This was known as the trans-earth injection. On 27 December, the craft re-entered the earth's atmosphere, travelling at 7 miles a second. It was the fastest return yet attempted. The capsule bounced twice like a stone on water, the planned method to slow it down. Deceleration reached 6g. High over the Pacific, communications were established with the recovery vessel USS *Yorktown* and the parachutes deployed. The capsule splashed down and turned upside down in the wind. Self-righting balloons activated and the module sat bobbing in the water. By dawn the recovery teams reached them. They were home from the moon.

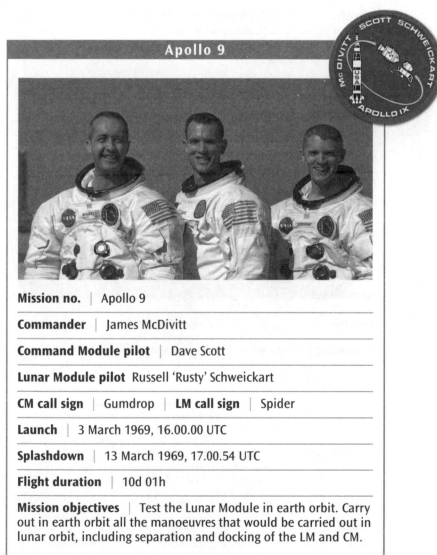

Apollo 9

Mission no.	Apollo 9
Commander	James McDivitt
Command Module pilot	Dave Scott
Lunar Module pilot	Russell 'Rusty' Schweickart
CM call sign Gumdrop	**LM call sign** Spider
Launch	3 March 1969, 16.00.00 UTC
Splashdown	13 March 1969, 17.00.54 UTC
Flight duration	10d 01h

Mission objectives | Test the Lunar Module in earth orbit. Carry out in earth orbit all the manoeuvres that would be carried out in lunar orbit, including separation and docking of the LM and CM.

For this and all subsequent Apollo missions, the crew were allowed to name their craft. The Command Module was dubbed *Gumdrop* and the Lunar Module was named *Spider*. The mission took place in earth orbit and lasted ten days. All its objectives were achieved. Russell Schweickart performed the first spacewalk. Schweickart and McDivitt uncoupled the Lunar Module and for six hours flew it 111 miles from the Command Module, after which David Scott performed a textbook docking. The mission ended after 151 orbits. Of the crew, only Scott would return to space, landing on the moon in Apollo 15.

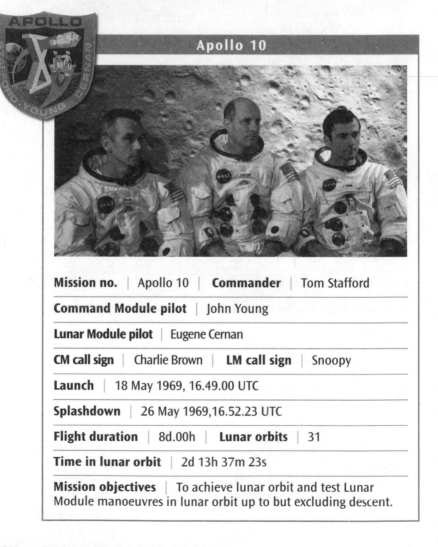

Apollo 10

Mission no.	Apollo 10	**Commander**	Tom Stafford

Command Module pilot | John Young

Lunar Module pilot | Eugene Cernan

CM call sign | Charlie Brown | **LM call sign** | Snoopy

Launch | 18 May 1969, 16.49.00 UTC

Splashdown | 26 May 1969,16.52.23 UTC

Flight duration | 8d.00h | **Lunar orbits** | 31

Time in lunar orbit | 2d 13h 37m 23s

Mission objectives | To achieve lunar orbit and test Lunar Module manoeuvres in lunar orbit up to but excluding descent.

Apollo 10 was the dress rehearsal for a lunar landing. On 22 May it safely entered lunar orbit and at 19.00.52 UTC Tom Stafford and Eugene Cernan climbed into the Lunar Module and undocked, leaving John Young alone in the Command Module. Both craft flew in formation across the far side of the moon. The Lunar Module's rockets were fired and *Snoopy* descended and went into an orbit that would take them within 16 kilometres of the moon's surface. Flying at 3730 miles per hour they surveyed landing sites for Apollo 11 and tested all of the module's functions short of landing. Only one mishap occurred, when the LM (pronounced 'lem') began to pitch and yaw round all its axes after a switch thrown by Cernan was inadvertently flipped back by Stafford. The gyrations were so severe that both pilots had difficulty seeing the controls. Stability was regained after 15 seconds. The Lunar Module flew across the moon's surface for 8 hours and 10 minutes before redocking with the Command Module. With Cernan and Stafford safely aboard, Command Module pilot Young turned *Charlie Brown* for home. They left the descent stage of *Snoopy* in lunar orbit and *Snoopy*'s ascent stage in a solar orbit, where it remains. *Charlie Brown* splashed down on 26 May 1969. On its way home, they transmitted the first colour TV pictures sent from space and set a record for the fastest manned vehicle, reaching a speed of 24,791 miles per hour.

Kennedy Space Center Deputy Director Albert Siepert (third row, left) points out highlights of the Apollo 10 liftoff to the King and Queen of Belgium.

The seven Apollo moon landing missions

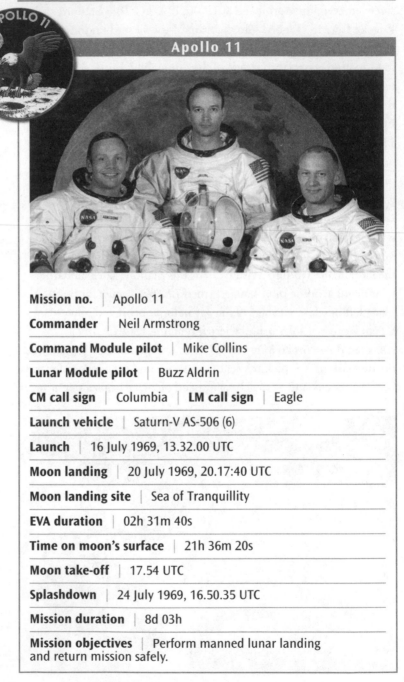

Apollo 11

Mission no.	Apollo 11
Commander	Neil Armstrong
Command Module pilot	Mike Collins
Lunar Module pilot	Buzz Aldrin
CM call sign	Columbia	**LM call sign**	Eagle
Launch vehicle	Saturn-V AS-506 (6)
Launch	16 July 1969, 13.32.00 UTC
Moon landing	20 July 1969, 20.17:40 UTC
Moon landing site	Sea of Tranquillity
EVA duration	02h 31m 40s
Time on moon's surface	21h 36m 20s
Moon take-off	17.54 UTC
Splashdown	24 July 1969, 16.50.35 UTC
Mission duration	8d 03h
Mission objectives	Perform manned lunar landing and return mission safely.

EAGLE: Twenty seconds and counting, T minus
fifteen seconds, guidance is internal. Twelve,
eleven, ten, nine, ignition sequence starts.
Six, five, four, three, two, one, zero. All
engines running. Lift-off, we have lift-off.
Thirty-two minutes past the hour, lift-off
on eleven . . .

Apollo 11 took off at 13.32 on 16 July 1969 and entered lunar orbit
on 19 July. It made 30 orbits, passing over the planned landing site
in the Sea of Tranquillity that had been recced by Ranger 8 and
the Surveyor 5 lander. On 20 July Neil Armstrong and Buzz Aldrin
entered the *Eagle* Lunar Module. After undocking from *Columbia*
the two craft flew in formation so that Collins, piloting *Columbia*,
could make a visual inspection of *Eagle*. Once Collins was satisfied
that all was well, Armstrong and Aldrin began their descent.

The descent would be controlled by the onboard Guidance and
Navigation System. For its time this computer was a miracle of design. But it was tiny. It had a fixed wired memory of 74 kilobytes and the equivalent of 2 kilobytes of 16-bit RAM, far less than the systems that control modern cars. As the descent began, Armstrong and Aldrin would be just over 25 miles above the moon's surface travelling backwards at 3600 miles per hour. The descent

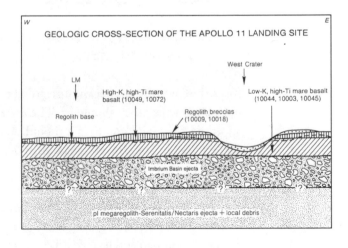

radar had to be able to see the ground, which meant the astronauts
would be looking towards the earth. The LM would travel in a
logarithmic curve, decelerating as it went. At 7000 feet the LM would
flip over so that Armstrong and Aldrin could see the moon and only
at 2000 feet would Armstrong be able to fly the LM by hand. Before

that the computations were too complex to be handled by anything other than the computer. The LM was cleared for descent. Radio communications were not good and Collins in the Command Module relayed orders to the LM.

> CAPCOM: *Eagle*, Houston. If you read, you are GO for powered descent. Over.
>
> COLUMBIA: *Eagle*, this is *Columbia*. They just gave you GO for powered descent.
>
> CAPCOM: *Columbia*, Houston. We've lost them on the high gain again. Would you please - we recommend they yaw right ten degrees and reacquire.
>
> COLUMBIA: *Eagle*, this is *Columbia*. You are GO for a PDI and they recommend you yaw right 10 degrees and try the high gain again.
>
> COLUMBIA: *Eagle*, you read *Columbia*?
>
> EAGLE: Roger, read you.
>
> COLUMBIA: OK.
>
> CAPCOM: *Eagle*, Houston. We read you now. You're GO for PDI.

Just before the descent began, Armstrong noticed that they were not exactly where they should be. The crater Maskelyne W had appeared 4 seconds later than it should. At the speed of more than 3000 miles per hour the error put them over two miles off course. The motors fired and the LM rolled over. The surface of the moon disappeared from view. Neither Armstrong nor Aldrin realized that they had left the LM rendezvous radar on. This was providing data that would only be used as they left the moon. The tiny computer went into overload and flashed a 1202 alarm Executive Overflow. It took Ground Control a long thirty seconds to decide that the mission could go on. The GO for continued descent came from the 26-year-old Guidance Officer (GUIDO), Steve Bales. After the mission Bales would say:

'It took us a long time. In the Control Center anything over three seconds in descent is too long. It took us ten or fifteen seconds.'

Along with the astronauts, Bales would earn the US Medal of Freedom for his actions. As the descent slowed and they passed 9000 feet, the LM rolled over again. Its feet pointed to the ground for landing. Armstrong and Aldrin could see the Sea of Tranquillity passing below them, but they had no idea where they were relative to the landing site. They could have been 2 miles or 20 miles adrift. Another alarm flashed on, a 1201. Guido Bales overrode it and called GO. They were at 2000 feet.

EAGLE: We're GO hang tight. We're GO two thousand feet.
CAPCOM: Roger.
EAGLE: Forty-seven degrees.
CAPCOM: *Eagle,* looking great. You're GO.

The speed of the LM came right down. Ahead, Armstrong could see a crater and rocks and a potential crash site. He had partial control of the flight. Aldrin was calling out their height. Armstrong headed for a gap in the rocks 'the size of a big house lot'.

EAGLE: Two hundred feet, four and a half down, five and a half down, one sixty, six and a half down, five and a half forwards, five per cent, quantity eight. Seventy-five feet down a half, six forwards.
CAPCOM: Sixty seconds [to fuel burn out].
EAGLE: Lights on.
EAGLE: We've got some dust. Sixty feet, down two and a half, two forward, that's good.
CAPCOM: Thirty seconds.
EAGLE: Contact light. That's good. Shutdown. OK engine stop, ACA out of détente. Descent engine command override off. Engine arm, off. Four One Three is in.
CAPCOM: We copy you down, *Eagle.*
EAGLE: Houston, Tranquility base here. The *Eagle* has landed.

The moon's first visitors from earth had arrived.

They spent the next six hours doing post-landing checks and preparing to step on to the moon. Aldrin took private communion and before he did so he transmitted this message:

This is the Lunar Module pilot. I'd like to take this opportunity to ask every person listening in, whoever and wherever they may be, to pause for a moment and contemplate the events of the last few hours and to give thanks in his or her own way.

At 2.56 UTC, 22 July 1969, six and a half hours after landing, Armstrong opened the hatch and stepped on to the moon.

I'm at the foot of the ladder. The LM footpads are only depressed in the surface about one or two inches, although the surface appears very fine-grained, as you get close to it, it's almost like powder. I'm going to step off the LM now. That's one small step for a man, one giant leap for mankind'.

Aldrin joined him:

Beautiful, beautiful, magnificent desolation.

The time and place

Time: 20.17.40 UTC
Landing coordinates: 0° 4' 25.69"N, 23° 28' 22.69"E
Geographical position: Sea of Tranquillity, the moon

They would spend two and a half hours walking outside. They discovered that movement was easy, 'even perhaps easier than the simulation'. They deployed the TV camera so that it could take panoramic views of the moon. Aldrin used the geologist's hammer to take soil samples and Armstrong photographed the Lunar Module so that engineers could use the photographs to assess how it had dealt with the rigours of landing. They planted the American flag and took an unscheduled phone call from President Nixon. They deployed the Early Apollo Scientific Experiments Package (EASEP).

Buzz Aldrin steps onto the moon.

This consisted of two packages, which contained a Laser Ranging Reflector, a Passive Seismic Experiment and a Solar Wind Collector. The Laser Ranging Reflector would be used to bounce back laser beams fired from the earth in a long-term experiment to measure the movements of the moon. Future missions would leave a more complex set of experiments known as the Apollo Lunar Surface Experiment Package (ALSEP). They also left memorabilia: an Apollo 1 mission badge, a disc bearing the names of heads of state and heads of NASA, Soviet medals commemorating cosmonauts Gagarin and Komarov, and a plaque bearing the inscription: 'Here Men from the Planet Earth First Set Foot Upon the Moon July 1969 AD. We Came in Peace for All Mankind.'

President Nixon has a post-splashdown talk with Apollo 11 Astronauts in the Mobile Quarantine Facility.

They re-entered the Lunar Module. They could smell gunpowder, the smell of the dust from the lunar regolith clinging to their suits. Then they slept. Before leaving the moon, they threw their back packs, their lunar overboots and a Hasselblad camera out of the module. At 17.45 UTC they disconnected from the descent stage and began the trip back into lunar orbit and a rendezvous with Collins. As they rose into the lunar sky Aldrin reported that he saw the American flag topple over in the downblast from the engine exhausts.

The docking with *Columbia* was tricky but successful. They jettisoned *Eagle* and, carrying 47.2 pounds of lunar rock, began the journey home. On their last night in space the three astronauts made a final broadcast to the world. Collins thanked all the men and women who had made the mission possible. Aldrin read from the Book of Psalms and Armstrong thanked NASA, the US government and the world.

Splashdown took place in the Pacific, 24 miles away from the recovery ship USS *Hornet*. The crew were immediately taken into a quarantine facility, through the armoured window of which they talked again to President Nixon, who joined the *Hornet* to meet them. They were home, brought 'safely back to earth'.

Apollo 12

Mission no. | Apollo 12

Commander | Pete Conrad

Command Module pilot | Dick Gordon

Lunar Module pilot | Al Bean

CM call sign | Yankee Clipper

LM call sign | Intrepid

Launch vehicle | Saturn-V AS-507

Launch | 14 November 1969, 16.22.00 UTC

Moon landing | 19 November 1969, 06.54.35 UTC

Moon landing site | Oceanus Procellarum (Sea of Storms)/ Mare Cognitum (Known Sea)

EVA duration | EVA 1: 3h 56m 03s | EVA 2: 3h 49m 15s Total EVA time: 7h 45m 18s

Time on moon's surface | 31h 31m

Splashdown | 24 November 1969, 20.58.24 UTC

Mission objectives | To perform manned lunar landing and return mission safely.

Apollo 12 took off in a storm – 36.5 seconds after launch the rocket was hit by lightning. The ensuing electrical faults caused a loss of

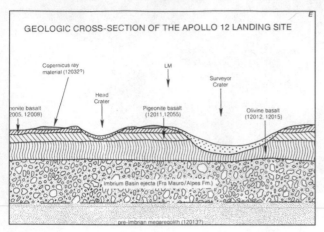

GEOLOGIC CROSS-SECTION OF THE APOLLO 12 LANDING SITE

the data feeds from the Command Module systems to Ground Control. The feeds were immediately restored but were garbled and incomprehensible. The mission was seconds away from an abort. John Aaron, EECOM officer, responsible for the onboard systems, remembered a similar incident in a simulation a year before and suggested:

'Try SCE to aux.'

This was an obscure command and for an instant neither Commander Pete Conrad nor anyone else could remember how to implement it. Al Bean suddenly remembered that a switch on his panel executed the command. He flipped it, coherent telemetry was re-established, and the mission was GO.

On 19 November the Lunar Module *Intrepid,* with Bean and Conrad aboard, undocked from the Command Module *Yankee Clipper* and began the descent to the moon. This was to be a precision landing using new Doppler effect radar techniques. The landing site was nicknamed 'Pete's Parking Lot' and it was about 400 yards from the unmanned Surveyor 3, *Snowman.* As Conrad made visual contact with the landing ground he exclaimed:

'There it is, right down the middle of the road.'

Worried that the planned landing site was not as favourable as had been thought, Conrad took manual control and landed about 200 yards from the marooned Surveyor. Five hours later Conrad left the LM to begin the first of two Extra-Vehicular Activities. As his feet touched the moon he exclaimed:

'Whoopee! Man, that may have been a small one for Neil but that's a long one for me.'

In so saying he won a bet with journalist Oriana Fallaci, who had insisted that Neil Armstrong's first words on the moon, 'That's one small step for a man . . .' had been scripted by NASA. Conrad bet her $500 that he could say what he wanted and gave her the script of what he would say. He won the bet but claims he never got the money.

Bean and Conrad's time on the moon was largely spent performing two EVAs and making jokes. At the beginning of the first walk, as Bean was deploying the colour TV camera, he pointed it at the sun, destroying the vidcom tube. In spite of Conrad's attempts to fix it by hitting it with the geological hammer it never worked again and there were no TV transmissions covering their mission.

The first walk lasted nearly four hours and in that time they deployed the Lunar Surface Experiment Package (LSEP) which contained equipment to measure solar wind, seismic activity, magnetic fields and the tenuous lunar atmosphere. One of the atmosphere detectors broke down after fourteen hours, its sampling having recorded only the gases vented from the astronauts' space suits.

The second walk's primary objective was to retrieve parts of Surveyor 3 for examination on earth. They closed the 200 yards to the marooned vehicle very fast and arrived exhausted. Conrad had planned to take a timed photograph of himself and Bean beside the Surveyor to fool people on earth into wondering who had held the camera. Unfortunately he mislaid the timer and aborted the joke.

Three hours and 49 minutes later they were back in the Lunar Module, having picked up the solar wind collector on the way. On the flight back to the CSM, Conrad gave Bean command of the *Intrepid*, giving him his only chance to fly the spacecraft. This was strictly against orders. Once reunited with Richard Gordon aboard *Yankee Clipper*, they jettisoned the LM, which crash-landed on the moon. The impact created an artificial earthquake that was measured by the seismic detectors deployed as part of the LSEP package. The reverberations lasted for more than an hour and caused one NASA engineer to declare that the moon 'rang like a bell'.

Splashdown on 24 November was rough. A camera dislodged and hit Bean on the head, knocking him out and giving him a cut that would need six stitches. On the moon the landing stage of *Intrepid* waits. The artist Forest 'Frosty' Myers claims that in one of its legs is an iridium-plated ceramic wafer with work by Andy Warhol, David Novron, Claus Oldenburg, John Chamberlain and himself etched into its surface. The chip is a tiny 3/4 × 1/2 × 1/4 inches and is called *Moon Museum*. Myers claims that he put it there with the help of an Apollo engineer at Grumman Corporation after NASA failed to give the piece official permission to be aboard the craft.

Apollo 13

Mission no.	Apollo 13
Commander	Jim Lovell
Command Module pilot	Jack Swigert
Lunar Module pilot	Fred Haise
CM call sign	Odyssey **LM call sign** Aquarius
Launch	11 April 1970, 19.13.00 UTC
Moon landing	Mission cancelled owing to onboard explosion
Splashdown	17 April 1970, 18.07.41 UTC
Mission duration	5d 22h
Mission objectives	To land and return safely. To perform geological and photographic surveys and to deploy Lunar Surface Experimental Package.

James Lovell, John Swigert and Fred Haise, the crew of Apollo 13, did not know, 321,859 miles into their mission, that their Command and Service Module (CSM), *Odyssey*, had a problem and that the problem had been with them since before take-off. A change in the electrical specification to the liquid oxygen tanks from 28 volts to 65 volts had damaged the no. 2 tank thermostat. This had led to undetected overheating, which in turn had burnt off the Teflon coating protecting the electrical cables in the tank. The accidents had turned the tank into a bomb. Any electrical activity in the damaged cable would detonate the liquid oxygen. Nobody knew this had happened, least of all the crew. Apollo 13 thundered on the launch pad at Cape Canaveral and achieved perfect lift-off.

The accidents had turned the tank into a bomb. Any electrical activity in the damaged cable would detonate the liquid oxygen.

Forty-six hours into the mission, Ground Control asked the Apollo crew to activate fans inside the no. 2 tank. Electricity pulsed along the damaged cable. Then, 321, 860 miles into the mission, the no.2 oxygen tank exploded, crippling *Odyssey*'s ability to make electricity and seriously jeopardizing its chances of getting home. It was unclear at this point what other damage the explosion had caused.

The Apollo 13 moon landing was aborted. Mission Control's priority now was to get the crippled vessel back to earth. Unable to generate its own electricity, the *Odyssey* had battery power for only ten hours. This was compounded by the immediate need to fire the main engine and put the vessel into FRT – Free Return Trajectory. This would enable *Odyssey* to use the moon's gravity to 'slingshot' it back to earth. Nobody knew whether it was safe to fire the main engine at all. Mission Control engineers were concerned that the engine had been damaged and that on ignition it too would explode. Just before *Odyssey* flew behind the moon and communication was lost, it was decided to attain FRT by using the engine on the now redundant Lunar Module, *Aquarius*. The manoeuvre was successful. *Odyssey* reappeared safely, made a second course correction, and she was on her way home.

Now the oxygen problem had to be tackled. It was running out. With no electricity-generating capability they could not scrub CO_2 from the cabin and if they could not do that they would suffocate. Again it was the Lunar Module *Aquarius* that came to the rescue. But there was a downside. The LM was equipped to support two men for two days, not three men for four. Eventually the oxygen problem was solved by modifying the lithium hydroxide CO_2 scrubbers on board *Aquarius* so that they would work on board *Odyssey*. This was a desperate action, using improvised equipment, on a tiny vehicle travelling in a near vacuum hundreds of thousands of miles from base.

Things got worse. The temperature inside *Odyssey* dropped, causing condensation throughout the vessel. This could result in short circuits and major electrical damage, which would happen when the electronics were turned on for re-entry. The final problem was that it was impossible to tell if the heat shield had been damaged. If it had, then *Odyssey* would simply burn up and disintegrate as she hit the earth's atmosphere.

Things were very tense inside the module and in Mission Control. The moon smiled on the three astronauts. *Odyssey* achieved safe re-entry and splashed down in the Pacific. The crew were safely picked up by the aircraft carrier USS *Iwo Jima*. *Aquarius*, the Lunar Module which had effectively acted as a lifeboat to *Odyssey*, never touched the moon, but it saved the lives of its crew. *Aquarius* was jettisoned over the Pacific and burnt up in the earth's atmosphere.

As a result of entering the Free Return Trajectory, the Apollo 13 mission altitude on the far side of the moon was 100 kilometres higher than any other Apollo mission. The mission is listed in the *Guinness Book of Records* as having the absolute altitude record for a manned spacecraft.

For the superstitious there are some odd coincidences surrounding the Apollo 13 mission. The number 13 is itself considered to be unlucky. The mission took off on 11.4.70 (digits add up to 13) at 13.13 Central Standard Time (CST), from Complex 39 (13 × 3). The explosion occurred at 19.13 CST on 13 April and a post-flight estimate of the cost of the damage was $13 million.

Apollo 14

Mission no. | Apollo 14

Commander | Al Shepard

Command Module pilot | Stu Roosa

Lunar Module pilot | Ed Mitchell

CM call sign | Kitty Hawk

LM call sign | Antares

Launch | 31 January 1971, 21.03.02 UTC

Moon landing | 5 February 1971, 09.18.11 UTC

Moon landing site | Fra Mauro

EVA duration | EVA 1: 04h 47m | EVA 2: 04h 34m
Total EVA time: 09h 22m

Time on moon's surface | 1d 09h 30m

Splashdown | 9 February 1971, 21.05.00 UTC

Mission duration | 9d 00h 01m

Mission objectives | To land and return safely. To perform geological and photographic surveys and to deploy Lunar Surface Experimental Package. Deploy and use Modular Equipment Transporter (a space wheelbarrow).

In the afternoon of 31 January 1971, the Apollo 14 mission was entering the last hours of countdown. The Apollo programme was unpopular with the government and the public were bored by it. Funding was being cut back. Two future missions had already been cancelled. The three astronauts going through the familiar ritual were aware that the reputation of NASA and the future of the Apollo programme were riding in their rocket. Between them they had logged less than 16 minutes in space. Those minutes belonged to the commander and self-made millionaire Alan Shepard. Lunar Module pilot Edgar Mitchell had never been in space. He had an interest in parapsychology and while in space planned to communicate telepathically with an ESP group in Florida. Command Module pilot Stuart Roosa had spent time with the US Forest Service and in his space suit pockets carried the seed of five tree types which he planned to distribute on return. The seeds, Loblolly Pine, Sycamore, Sweetgum, Redwood and Douglas Fir, made it back from space and grew into lunar trees that can be found all over the world. One, a pine, grows in the garden of the White House.

The three men, called by Apollo 12 commander Pete Conrad 'the rookies-only crew', took off at 21.03 UTC and two hours later hit their first problem. In earth orbit the Lunar Module had to be separated, turned round and redocked with the Command Module. *Antares* refused to dock with the CM, *Kitty Hawk*. This was solved by firing the *Kitty Hawk*'s thrusters to hold it against *Antares* and retracting the docking probe. This worked and the docking latches snapped into place. The next problem came as the LM began its powered descent. An abort switch developed an intermittent problem, causing it to activate. They tried tapping the panel next to the switch, a solution which had limited success. In the end the computer was reprogrammed manually, with Mitchell entering the new strings of code until they were less than 2 miles above the moon's surface. Luckily it worked. The craft came to rest on the moon, one foot in a small crater, which tilted it to 7°. They were closer to their planned landing spot than any mission so far. Shepard's comment was:

'We made a good landing.'

Six hours later, in the early morning of 5 February, he stepped on to the surface of the moon, an experience he had waited ten years to achieve. He was similarly laconic:

'It's been a long way, but we're here.'

For the next 4 hours and 47 minutes they deployed a laser retro-reflector to augment the one left by Apollo 11, a solar wind collector, a seismometer, a solar particle detector and equipment to study the lunar atmosphere. They also deployed a grenade launcher which was to be used as part of the seismology programme, and they detonated twenty small charges as part of the investigation into the nature of the ejecta surface they were on. There followed a period of rest.

The next walk, begun about 18 hours after the first, called for the Modular Equipment Transporter (or 'lunar rickshaw') to be dragged over half a mile to the edge of the Cone crater. The walk was hard and map reading proved difficult. The astronauts wore rigid space suits; their heart rates reached 150. At one point they were confused as to where they were. Mitchell exclaimed:

'Our positions are all in doubt.'

The excursion lasted 4 hours and 34 minutes. Samples were gathered, movie-camera footage was shot and magnetometer recordings were made. Before re-entering the LM, Shepard used a golf club he had smuggled aboard to hit a ball for what he called miles and miles – in fact a few hundred yards. Mitchell used the solar wind collector support as a javelin.

The return to earth was uneventful and the splashdown less than a thousand metres from the planned point. Mitchell and Shepard had temporarily saved NASA's reputation. Although they were criticized by geologists for poor in-field techniques, they had successfully dealt with more difficult docking and landing problems than those encountered by any other Apollo manned mission. They achieved many firsts, including the first colour TV transmissions from the moon's surface. The experiments they conducted produced fascinating data about the blanket covering Imbrium. The equipment they left behind recorded the impact of a large 1.2-ton meteorite, and also a small earthquake from which the supra-thermal ion detector detected rarefied gases from deep under the surface that may have contained water vapour.

Apollo 15

Mission no. | Apollo 15

Commander | Dave Scott

Command Module pilot | Al Worden

Lunar Module pilot | Jim Irwin

CM call sign | Endeavor | **LM call sign** | Falcon

Launch | 26 July 1971, 13.34.00 UTC

Moon landing | 30 July 1971, 22.16.29 UTC

Moon landing site | Hadley Rille

EVA duration | LM stand up: 00h 33m (commander opened Lunar Module hatch and looked out)
EVA 1: 06h 32m | EVA 2: 07h 12m | EVA 3: 04h 49m | Total EVA time: 18h 33m

Time on moon's surface | 2d 18h 54m

Lunar Roving Vehicle | LRV-1

Splashdown | 7 August 1971, 20 45 53 UTC

Mission duration | 12d 07h 11m

Mission objectives | To land and return safely. To perform geological and photographic surveys and to deploy Lunar Surface Experimental Package. Deploy and use the Lunar Roving Vehicle.

The Apollo mission moon badges were inspired by mythology, patriotism and wishful thinking. The map shows the sites of landings made by the American Apollo and Soviet Luna missions – the concrete achievements of the space race.

Before the Enlightenment, science and magic were two sides of the same coin. Science and the occult both drew on the power of the moon. The beautiful 'Anatomy of Man and Woman' (**right**) from the fifteenth-century Les Très Riches Heures du Duc de Berry shows the anatomical influence of astrological signs. The moon's sign cancer is on the chest.

The character of Luna is used in both the fifteenth-century calendar illustration (**below left**) and the tarot card showing Luna and her sign, Cancer the crab (**below right**).

The advent of scientific thought has not dulled our desire to understand the moon's power. The Luna tarot card (**above left**) shows the frightening effect that Diana had on her followers. The astrologers in the Luna tarot card (**above right**) are trying to measure the moon, while the American Fate Magazine from 1954 displays a touching faith in modern pseudo-scientific thinking about the moon.

TRUE STORIES OF THE STRANGE AND THE UNKNOWN

FATE
MAGAZINE

February 1954 35¢

THE MOON AND YOU

SCIENCE VIEWS THE FAITH CURE

DREAMS FOR PROFIT

What today we would consider occult works were published before the eighteenth century as scientific treatises. The picture of Hermes Trismegistus (above) holding a scientific instrument and standing beside the primal sources of the sun and the moon was originally published in *Chemisches Lustgärten* (*The Alchemical Pleasure Garden*), which became one of the most important works on alchemy. It was written by the physician Daniel Stolcius and eventually published as *Viridium Chymicum* (*The Encyclopedia of Alchemy*).

The meaning of alchemical imagery has been obscured by time. The emblem from the fifteenth-century *Alchemical Receipts* (above right) shows a powerful combination of alchemical emblems including a toad, the sun and moon, a serpent and a phoenix-like bird. The illustration from the sixteenth-century *Splendor Solis* (*The Splendour of the Sun*) (right) shows one of twenty-two plates describing the death and rebirth of the king. He is in a flask, suggesting the distillation of his spirit into the perfect state necessary for the epitome of the divine majesty of kings. There are only twenty copies of *Splendor Solis* in existence.

The meaning of this complex emblem from the *Speculum Philosophia* (opposite) is incomprehensible to most modern eyes, though would have been clear to eighteenth-century scientists and alchemists such as Sir Isaac Newton.

Mortuis Corporibg remanent Spiritus soluti, Igitur corporibus morte
vestitij illa equitatis cum Salie et cum ☉ ☽ lumine et   10
stellis fixeis.

Caput Corui dicitz

Laton dicitz ista figu-
ra idest principium
artij qui in vase
apparet niger, et
est principium
corruptionis.

Lapidis
ermetij

figura nigra superius est prima materia, quando ponitz in vase
ad ignem, fit ita nigra ascendendo gradatim ad albedinem per
scalam digestionis, et per gradus Ignis decoquendos.

Diei

The moon has exerted a powerful effect on modern culture. The worlds of literature, music and film have all felt its influence as can be seen here in the illustration from The Adventures of Baron Munchausen (**left**) by Eric Raspe, 1850; the cover to the sheet music for 'Blue Moon' by Rogers and Hart, and the three very different fantasy films: Cat-Women of the Moon, 2001: A Space Odyssey (**opposite, top**), and the animation A Grand Day Out (**opposite, bottom**). In 2001, director Stanley Kubrick achieved one of the most accurate descriptions of the moon ever seen on celluloid.

Two science-fiction attempts to describe the future colonization and exploitation of the moon. The machine in the picture below is an excavator designed to recover lunar minerals including the holy grail of power sources, Helium-3. Such work could lead to the destruction of the moon's surface and its near perfect vacuum environment. The City on the Moon *by Chris Butler* (**right**) shows how a crater might be used to build a city partly shaded from the fierce glare of the sun.

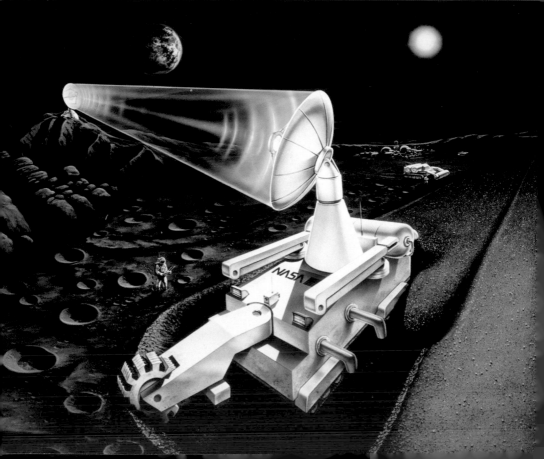

The all-Air Force crew that waited for liftoff in the Command and Service Module of Apollo 15 had been the back-up crew to the all-Navy crew of Apollo 12. They carried the heaviest payload, had the most focused scientific agenda and would stay longer and travel further than any of the previous Apollo missions. The actions of the crew once they returned would also prove the most contentious.

Apollo 15 was the first J-type mission, an extended scientific exploration. Apollo 11 had been designated a G-type mission, a landing; and Apollos 12, 13 and 14 had been H-type missions, the exploration of other landing sites. I-type missions had been planned – unmanned orbital investigation – but, under pressure to deliver, NASA had cancelled them.

The J-type missions carried the heaviest payloads yet sent to the moon. The equipment included a Lunar Roving Vehicle (LRV) and a Scientific Instrument Module (SIM), which added significant extra weight. The LRV had been built by Boeing and had an unladen weight of 500 pounds. It could carry about 900 pounds of personnel, equipment and samples, could travel at about 11 miles per hour and had a there-and-back range of 40 miles. The Scientific Equipment Module on Apollo 15 included photogrammetric cameras, laser altimeters, spectrometers and a Particles and Fields satellite that could be launched in lunar orbit. The satellite would collect data on interplanetary magnetic fields and other such phenomena.

The crew had received extensive training in geological observation. The space suits were much more flexible than on the previous missions.

The mission landing site was near to Mount Hadley, where they would be able to investigate terrain that was thought to have been formed by volcanic activity. On Friday, 30 July, the LM began its descent to the surface. Above them in the Command Module, Worden would be busy making photographic examination of the ground beneath his orbiting spacecraft.

The *Falcon* landed at 22.16 UTC. Two hours later work had begun. First Scott made a visual examination of the terrain through the top hatch of the Lunar Module (in flight this hatch allowed access to the Command Module). After this preliminary recce,

Scott left the craft. His first words on the moon were:

'Man must explore and this is exploration at its greatest.'

He and Irwin deployed the lunar rover and became the first men to drive on the moon's surface. During the first EVA they drove in the LRV to collect rock samples and deployed the ALSEP.

The second EVA began nearly 24 hours later. Most of the time was spent taking rock samples. These proved to be very interesting. One sample was of a greenish material that was discovered to be magnesium-rich volcanic glass. This had come from the moon's mantle and had been ejected with such force that it was very pure. They also found a piece of rock that had come from the original lunar crust. This rock, catalogued as number 15415, is now called the Genesis Rock. It is 4.5 billion years old and was formed when our solar system had existed for a mere 100 million years.

During the third EVA more geological work was undertaken, adding to our knowledge of the volcanic activity that took place early in the moon's life.

Before they left, Scott performed a small demonstration of gravity. He dropped a feather and a hammer simultaneously. They landed at the same time, confirming the theories about gravity first propounded by Galileo.

The crew returned to earth. As the LM took off, Collins and Mitchell played the US Air Force song 'Wild Blue Yonder'. As well as the scientific equipment they left behind a small star known as 'The Fallen Astronaut'. It commemorated the American and Russian astronauts who had died in the pioneering days of space exploration. Before they left lunar orbit they released the small Particles and Fields satellite. It would send data back to earth for the next six months. On the way home Worden left the CM to perform a 39-minute spacewalk to recover the film with his fly-by mapping photographs.

Splashdown took place at 20.45 UTC in the Pacific and with it the mission, which had lasted just over twelve days, came to an end.

The aftermath was marred by a scandal. The astronauts had smuggled stamps and commemorative envelopes aboard their spacecraft. The plan to sell them fell through and the three men were reprimanded by NASA.

Apollo 16

Mission no. | Apollo 16

Commander | John Young

Command Module pilot | Ken Mattingly

Lunar Module pilot | Charlie Duke

CM call sign | Casper

LM call sign | Orion

Launch | 16 April 1972, 17.54.00 UTC

Moon landing | 21 April 1972, 02.23.35 UTC

Moon landing site | Descartes Highlands

EVA duration | EVA 1: 07h 11m | EVA 2: 07h 23m
EVA 3: 05h 40m | Total EVA time: 20h 14m

Time on moon's surface | 2d 23h 02m

Splashdown | 27 April 1972, 19.45.05 UTC

Lunar Roving Vehicle | LRV-2

Mission duration | 11d 01h 51m

Mission objectives | To land and return safely. To perform geological and photographic surveys and to deploy Lunar Surface Experimental Package. Deploy and use the Lunar Roving Vehicle.

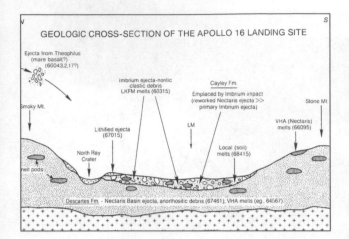

GEOLOGIC CROSS-SECTION OF THE APOLLO 16 LANDING SITE

Orion, the Lunar Module of Apollo 16, landed on 21 April 1972 at 02.23.35 UTC. It was six hours late. After undocking, the spacecraft had begun to shake very violently. The crew were ordered to continue in orbit with the Command Module, *Casper*, until the problem could be fixed or until the mission was aborted. With the problem solved, the mission continued. Young and Duke performed three long lunar walks. Like all previous mission commanders, Young had his shot at first words on the moon. His were:

'There you are, our mysterious and unknown Descartes Highland Plains; Apollo 16 is going to change your image.'

Their mission was to explore what was thought to be volcanic rock in the central Highlands near the large crater Descartes. This material turned out to be not volcanic but ejecta, rock thrown out and vaporized in the enormous temperatures generated by meteors hitting the moon.

On their first walk the two crewmen deployed the ALSEP. An important experiment which entailed placing thermal couples in boreholes had to be abandoned when the wires from the probes became wound round Young's legs and were pulled from their mounting. The discovery that the rocks they were finding did not appear to be volcanic caused the geologists at Mission Control to request the retrieval of a rock that appeared to glint and which therefore might have been thrown from the magma under the moon's surface. This rock weighed 26.5 pounds and was the largest rock ever brought back from the moon. It was named Big Mulley after the senior mission geologist, William Muehlberger. Young concluded the first EVA by driving the LRV at top speed round a circuit, braking and accelerating as hard as he could. He travelled 5000 yards.

Lunar Module pilot Charlie Duke collecting lunar samples. He is standing on the edge of Plum crater and the lunar rover can be seen in the background.

During the second EVA Young and Duke gathered more rock samples. One had red marks on it which were at first thought to be caused by water but which later turned out to be the results of human contamination. Again, no volcanic rocks were recovered. This excursion lasted 7 hours 23 minutes, in which time the LRV travelled over 12,000 yards.

In the final EVA they took samples from the largest rock yet encountered on the moon, the House Rock. The EVA ended after 5 hours and 40 minutes. The mission itself ended a day early and after releasing the Particles and Fields satellite (which malfunctioned) CM *Casper* returned to earth, pausing en route to allow Mattingly to carry out a spacewalk to recover the film he had shot while orbiting the moon in the Command Module, and to expose a colony of bacteria to the conditions of deep space.

The crew of Apollo 16 were given a modified version of the diet their predecessors in the earlier Apollo missions had eaten. Orange juice formed a large part of their intake and they experienced few in-flight health problems, though they suffered more from flatulence.

Apollo 16, like Apollo 15, collected detailed data about the moon's magnetic fields and the phenomena known as 'mascons' – anomalies in those magnetic fields. The lack of volcanic rock caused geologists to reconsider their ideas about the formation of the Cayley Plains.

Apollo 17

Mission no. | Apollo 17

Commander | Gene Cernan

Command Module pilot | Ron Evans

Lunar Module pilot | Jack Schmitt

CM call sign | America

LM call sign | Challenger

Launch | 7 December 1972, 05.33.00 UTC

Moon landing | 11 December 1972, 02.23.35 UTC

Moon landing site | Taurus–Littrow

EVA duration | EVA 1: 07h 11m | EVA 2: 07h 36m
EVA 3: 07h 15m | Total EVA time: 22h 02m

Time on moon's surface | 3d 02h 59m

Splashdown | 19 December 1972,19.24.59 UTC

Lunar Roving Vehicle | LRV-3

Mission duration | 12d 13h 51m

Mission objectives | To land and return safely. To perform geological and photographic surveys and to deploy Lunar Surface Experimental Package. Deploy and use Lunar Roving Vehicle.

This mission was the last in the Apollo series. While Eugene Cernan and Ronald E. Evans were military men, Harrison 'Jack' Schmitt was a trained geologist. He had joined the astronaut programme in 1965, chosen as one of six specialists from a field of about 1000 applicants. Part of Schmitt's training had been in the handling of high-performance aircraft. Schmitt could fly the LM and confirmed NASA's opinion that, if properly trained, anyone could fly it.

Apollo 17 was the only mission in which the Saturn rocket was launched at night. Its trail could be seen 700 miles away and was so bright that it brought fish off Cape Cod to the surface, fooled into thinking it was the sun.

On board were five mice who were destined to take part in an experiment to test the effects of radiation on the nervous system. Much of the equipment carried in the hold was new. The LM ALSEP package included two gravimeters to measure lunar gravity, one of which would be deployed on the LRV to measure local gravity at the sampling sites. There were also a Lunar Regolith Electrical Characteristics Sensor and a neutron probe to study the effects of cosmic rays on the regolith. The Command Module carried a Particles and Fields satellite, a far ultra-violet spectrometer to examine the possibility of hydrogen atoms near the surface of the moon, an infra-red radiometer to measure surface temperature, as well as other equipment for measuring the regolith. This would give CM pilot Evans a busy time as he orbited the moon.

The destination was the Taurus–Littrow area of the moon. Landfall was made on 11 December 1972 at 02.23.35 UTC. They were a mere 330 yards short of the target and had two minutes of fuel left in the tank.

Cernan and Schmitt spent three days on the moon and made three seven-hour excursions. Thanks to Schmitt, the work they did was more focused and detailed than that of any of the other Apollo landings. Previous astronauts had tended to despise the geology. They preferred flying the mission and tearing about the moon's surface in the lunar rover. On Apollo 12 Pete Conrad took to referring to the rocks as just 'stuff'. Schmitt knew that he was looking at evidence of what the solar system had been like nearly 4 billion years ago. Geological evidence that had long been obliterated on the surface of the earth was, on the moon, his for the taking. It was

Schmitt too who took the most famous of all the lunar photographs: the earth as seen from the moon.

Schmitt and Cernan also set up three of the most interesting experiments performed on the surface of the moon. They deployed an ejecta and meteoroid detector which could measure the size, speed, mass and direction of meteorites, giving data which would contribute to understanding the formation of the lunar regolith. They left charges which would be detonated after take-off, the shockwaves of which would be recorded on geophones and would give evidence as to the nature of the moon's crust to a depth of over a mile. They left equipment which would monitor the lunar atmosphere. This revealed that the moon did have a very thin atmosphere consisting of helium and argon, which, having a mass of 66,000 pounds, would be enough to fill the Albert Hall at earth atmosphere densities. Other elements were discovered by terrestrial observers in the 1980s.

Before setting off on their first excursion, Schmitt gave a detailed and informed description of the landing site, far more insightful than had been achieved by his predecessors. The first EVA was used to deploy the ALSEP and to take a short drive to the crater Steno. During this trip the geological hammer attached to Cernan's lower leg broke part of the wheel guard on the lunar rover. The crew were sprayed by the potentially hazardous and very fine moon dust thrown up by the wheels. The damage was repaired using duct tape and a heavy map.

The objective of the second walk was to collect rock samples from the south Massif. They covered most of the distance in the rover and Schmitt spotted and collected one of the oldest rocks yet found – a piece of pure olivine nearly as old as the Genesis Rock picked up by the Apollo 15 astronauts David Scott and James Irwin. They also collected ejecta thrown up at the creation of the moon's latest large impact crater, Tycho. This turned out to be a mere 109 million years old. On earth, had they been looking, dinosaurs would have seen the impact that created it.

They then examined the crater Shorty, setting up an antenna so that live TV pictures of the expedition could be beamed back to earth. Schmitt caused a sensation by discovering orange soil, which could have been caused by oxidation and which, if this was true, would point to the presence of water under the moon's crust. In fact it turned out to be volcanic glass shot through the crust 4.5 billion years before. This material had been covered in lava and then revealed in the impact which formed Shorty.

The third and last EVA took them to a huge 10-metre-long rock which had broken into five pieces. It lay at the end of a visible trail which indicated that it had rolled hundreds of metres. Schmitt took his time to survey the rock and then declared that it was in fact two different types of rock welded together in the colossal heat caused by the impacting meteor.

The astronauts collected a total of 243.65 pounds of moon rock, among which was the Goodwill Rock, which was to be sliced up and given to the students of seventy countries. They left behind a plaque saying:

> *Here Man completed his first exploration of the moon*
> *December 1972 AD. May the spirit of peace in which we came*
> *be reflected in the lives of all mankind.*

After docking with the CM *America* the crew of three orbited the moon for two days. They observed the strange and beautiful phenomenon of rays of sunlight illuminating tiny particles of dust in the lunar dawn.

At 23.35 on 16 December they fired the Service Propulsion System engine for the last time and entered the trajectory which would take them safely and sadly home. Cernan delivered one last message to the world:

> *America's challenge has forged Man's destiny of tomorrow.*
> *And as we leave the Moon at Taurus–Littrow we leave as we*
> *came and, God willing, as we shall return with peace and*
> *hope for all mankind. Godspeed the crew of Apollo 17.*

His last, unrecorded and possibly apocryphal, words to Schmitt as he blasted off from the moon's surface were:

'OK, Jack, let's get this mother outta here.'

Apollo 11

Neil Alden Armstrong

US Navy pilot | 1st man on the moon | Commander
Born 5 August 1930 | Time in space: 8d 14h 12m

A Navy pilot with combat experience, Armstrong was selected for astronaut training in September 1962. He flew with David Scott on the Gemini 8 mission which had to be aborted after problems with the manoeuvring equipment. Armstrong and Scott were criticized for their handling of the problem but later exonerated. Flight Director Gene Krantz declared:

'The crew reacted the way they were trained; they acted wrong because we trained them wrong.'

While training for the Apollo missions Armstrong was at the controls of a Lunar Landing Training Vehicle when it began to malfunction. Armstrong ejected with half a second to spare.

Armstrong became the first man on the moon. He believed that Apollo 11 had only a 50 per cent chance of success. He said:

'I was elated, ecstatic and extremely surprised that we were successful.'

After the mission he never again flew in space. He continued to work with NASA and in the civilian world he accepted appointments in, among other things, the aerospace and oil industries.

Edwin Eugene 'Buzz' Aldrin

Colonel, US Air Force | 2nd man on the moon
Lunar Module pilot | Born 20 January 1930
Time in space: 12d 1h 52m

Aldrin's sister could not pronounce the word 'brother', eliding it to 'buzzer'. His family name then became 'Buzz', which Aldrin took as his legal name in 1988.

Aldrin was an Air Force pilot with combat experience. He attended the Massachusetts Institute of Technology and gained a Ph.D. His doctoral thesis was entitled *Line of Sight Guidance*

Techniques for Manned Orbital Rendezvous. The mechanics and techniques of rendezvous became something of an obsession for Aldrin. He was selected for astronaut training in October 1963. He helped devise new training techniques for Extra-Vehicular Activities, introducing neutral buoyancy underwater training to simulate weightlessness. As the Lunar Module pilot of Apollo 11 he became the second man to step on the moon.

After retiring from active service in 1972 he suffered from depression and alcoholism; he was greatly helped to beat these two illnesses by his wife, Lois Aldrin.

Aldrin was publicly confronted by the Apollo hoax theorist Bart Sibrel, who called him 'a coward, a liar and a thief'. Aldrin punched Sibrel in the face. Aldrin has appeared many times on television, sits on several charitable committees, is an inductee of the Astronaut Hall of Fame and in 2003 received the Humanitarian Award from the children's charity Variety. The citation reads 'given to an individual who has shown understanding empathy and devotion to mankind'. The popular cartoon character Buzz Lightyear is named after him.

Apollo 12

Charles 'Pete' Conrad, Jr
Captain, US Navy | *3rd man on the moon* | *Commander*
Born 2 June 1930; died 8 July 1999
Time in space: 49d 3h 38m

Conrad was a Navy pilot and instructor. His first application for astronaut training did not go well. He rebelled against the invasive medical and psychological tests. When asked what a Rorschach inkblot meant to him he paused and then said it was upside down. He was rejected as being unsuitable for long-duration flight.

Alan Shepard persuaded Conrad to reapply and in September 1962 he was successful. He piloted Gemini 5 and set a space endurance record of eight days. He went on to command Apollo 12, after which he flew in Skylab 2. Conrad retired from the Navy and NASA in 1973, after which he spent many years with McDonnell

Douglas. On Valentine's Day 1996 he set another endurance record as part of the crew of Cable TV magnate Bill Daniels's Lear jet that flew round the world non-stop in 49 hours 26 minutes and 8 seconds. Conrad was killed in a motorcycle accident; the place where he died is called Ojai, a Native American word for 'moon'.

His motto was: 'If you can't be good, be colourful.'

Alan LaVern Bean

Captain, US Navy | 4th man on the moon
Lunar Module pilot | Born 15 March 1932
Time in space: 69d 15h 45m

Alan Bean was a Navy pilot and a test pilot. He was selected for astronaut training in 1963. After his Apollo mission Bean flew as commander of Skylab 3. In 1981 he resigned from NASA to devote his time to painting. He says that he has seen things that no other artist has seen and it is his mission now to depict those things. He argues that as an artist he can bring colour to the moon, something that a scientist would not be able to do. He uses moon dust from his space suit in his paintings and among his brushes is the hammer that he used on the moon. Bean now uses this as part of his painting technique. His paintbrushes bearing his signature can fetch several hundred dollars. His works sell as originals and prints, at prices that range from a few to several thousand dollars.

Apollo 14

Alan 'Al' Bartlett Shepard, Jr

Rear Admiral, US Navy | 5th man on the moon
Commander | Born 18 November 1923; died 21 July 1998
Time in space: 9d 00h 57m

Shepard was a Navy pilot who served in the Second World War. In 1959 he was invited by NASA to apply for the first manned space flight programme. On 5 May 1961 he piloted the Freedom 7 mission, which made him the second man and first American to go into space. Before takeoff he said: 'Please, dear God, don't let me fuck up,' and this has since become the astronaut's prayer. In 1964 he was diagnosed with Ménière's disease, a condition of the inner ear the symptoms of which are disorientation, nausea and dizziness. Shepard was rendered unable to fly until 1969, when the problem was solved by surgery. During that period he served as the Chief of the Astronaut Office and also became a millionaire. In 1969 he commanded Apollo 14. On leaving NASA he pursued his highly successful business interests through his company Seven Fourteen Enterprises. He co-wrote *Moon Shot: The Inner Story of America's Race to the Moon*, which was adapted for TV in 1994. Shepard died of leukaemia on 21 July 1998. He was a descendant of *Mayflower* passenger Richard Warren.

Edgar 'Ed' Dean Mitchell

Captain, US Navy | 6th man on the moon
Lunar Module pilot | Born 17 September 1930
Time in space: 9d 00h 01m

A Navy pilot and devotee of the paranormal, Mitchell was selected for astronaut training in 1966. Apollo 14 was his only space mission. While in space he carried out unofficial extra-sensory perception experiments. He resigned from NASA in 1972 and in 1973 he founded the Institute of Neotic Sciences and wrote *Psychic Exploration: A Challenge for Science* and *The Way of the Explorer: An Apollo Astronaut's Journey through the Material and the Mystical Worlds*. He believes that UFOs are visitors from other planets and that there are world conspiracies to deny this. 'We know they are real, now the question is where they come from.' He believes he had a renal carcinoma

> *'The presence of divinity became almost palpable and I knew that life in the universe was not just an accident based on random processes.'*

that was healed by a teenage remote healer calling himself Adam Dreamhealer. Mitchell said that while in space 'The presence of divinity became almost palpable and I knew that life in the universe was not just an accident based on random processes.'

Apollo 15

David Randolph Scott

Colonel, US Air Force | 7th man on the moon
Command Module pilot | Born 6 June 1932
Time in space: 22d 18h 53m

Scott was an Air Force pilot and test pilot. He was selected for astronaut training in October 1963. He flew in Gemini 8 with Neil Armstrong and performed the first space docking. As Command Module pilot for Apollo 9 he completed a lunar orbit rendezvous and docking simulation. He next flew in Apollo 15 as commander. On retiring from NASA he became a media consultant. He commentated for the BBC coverage of the first space shuttle and was consultant on the film *Apollo 13* and the TV mini-series *From the Earth to the Moon*. He collaborated with the Soviet cosmonaut Alexei Leonov on the book *Two Sides of the Moon: Our Story of the Cold War Space Race*.

James Benson Irwin

Colonel, US Air Force | 8th man on the moon
Lunar Module pilot | Born 17 May 1930; died 8 August 1991
Time in space: 12d 7h 12m

An Air Force pilot and test pilot, Irwin was awarded the Distinguished Service Medal. Selected for astronaut training in 1966 he flew as Lunar Module pilot in Apollo 15. He retired from the Air Force and in 1972 co-founded an evangelical Christian ministry called High Flight. High Flight operates religious retreats and tours of the Holy Land. Irwin led several unsuccessful attempts to find Noah's Ark on Ararat. He said of these attempts that 'it is easier to walk on the moon. I've done all I possibly can but the Ark continues to elude us.' He wrote several books, *To Rule the Night, More Than Earthlings, More Than an Ark: Spiritual Lessons Learned While Searching for Noah's Ark* and *Destination Moon*. He died of a heart attack in 1991. Of his moon mission he said: 'I felt the power of God as I have never felt it before.'

Selected for astronaut training in 1966 he flew as Lunar Module pilot in Apollo 15.

Apollo 16

Charles Moss Duke, Jr
Brigadier General, US Air Force | 10th and youngest man to walk on the moon | Lunar Module pilot
Born 3 October 1935 | Time in space: 11d 1h 51m

An Air Force pilot, Duke joined NASA in April 1966. His voice is known throughout the world as the CAPCOM for Apollo 11. German measles prevented him from flying in Apollo 13. He flew as Lunar Module pilot in Apollo 16. After the mission and because of his experiences in space he became a Christian.

John Watts Young
Captain, US Navy | 9th man to walk on the moon
Command Module pilot | Born 24 September 1930
Time in space: 34d 19h 39m

Young was a Navy pilot and test pilot. He joined NASA in 1962 and started on a career in which he would become one of NASA's most experienced pilots. He got off to a bad start when as a member of the first manned space flight crew of Gemini 3 he smuggled a corned-beef sandwich on board. Although his career suffered a setback, he was not grounded. He flew as Command Module pilot of Apollo 10 and commander of Apollo 16. As back-up commander of the nearly disastrous Apollo 13 he played a crucial role in getting the stricken spacecraft back to earth. While preparing for the Apollo 16 mission he discovered an interest in geology. Young stayed with NASA for forty-two years. He retired in 2004, aged seventy-nine. He had flown more spacecraft than any other astronaut. He once stated: 'NASA is not about the "adventure of human space exploration". We are in the deadly serious business of saving the species.'

He once stated: 'NASA is not about the "adventure of human space exploration". We are in the deadly serious business of saving the species.'

Apollo 17

Eugene Andrew Cernan

Captain, US Navy | *11th and last man to walk on the moon*
Commander | *Born 14 March 1934*
Time in space: 23d 14h 15m

A Navy pilot, Cernan joined NASA in 1963. He flew three times into space, first with Gemini 9, then Apollo 10 and finally as commander of Apollo 17. Cernan is currently the last Apollo astronaut to have stood on the moon. On leaving NASA he wrote *The Last Man on the Moon*.

Harrison 'Jack' Schmitt

Geologist | *12th man to walk on the moon*
Lunar Module pilot | *Born 3 July 1935*
Time in space: 12d 13h 2m

Jack Schmitt was the first moon walker to have no military background. He trained as a geologist, taking a B.Sc. in Science from Caltech and a Ph.D. in Geology from Harvard. Before joining NASA he worked in the US Geological Survey Astrogeology Department developing field techniques for Apollo crews. He helped examine lunar samples, learnt to fly and trained astronauts in basic field techniques. After he flew Apollo 17, he resigned from NASA to stand for the Senate, representing New Mexico, which he did for one term. He is president and founder of the Interlune Intermars Initiative and believes we should return to the moon and exploit its natural resources, including Helium-3.

The Future

And now, in the twenty-first century, the moon race is on again. We have big plans for the moon, plans that will change it for ever. We are in the grip of a new moon madness. Tourist trips are being offered at $100 million dollars a return ticket and there are already buyers. We are poised to spend ever greater sums on its exploitation, money that might be better spent improving the living conditions of everyone on earth.

Whatever the outcome of the next phases of lunar exploration, we will never lose our fascination with the moon. Our understanding of it has come a long way since the first primitive astronomer carved moon phases on an eagle bone over 30,000 years ago. Since then we have spent billions of pounds and as many man hours trying to increase our understanding of what the moon is, how it functions and how it was formed. The moon is becoming as familiar as our own back yards.

As the moon becomes more and more familiar, we must be careful not to treat it with greater and greater contempt.

Earthrise seen from Apollo 8.

Chapter 6

Magic

The Occult, Astrology, Alchemy, Prophecy, Fortune-Telling, Spells and Superstition

The Occult

The occult is the other side of science. Where the scientist looks at and tries to discover what is there, the occultist looks below the surface for a hidden and greater truth. Practitioners of the occult arts deal in astrology, alchemy, magic, witchcraft, prophecy, fortune-telling and plain superstition. The occult world has its roots deep in prehistory. It is inhabited by spirits, ghosts, angels and demons. Occult laws are enshrined in ancient texts that are secret, coded, obscure and, if they exist at all, only partly understood. The occult can be a benign force that seeks earthly wisdom and spiritual enlightenment; it can also be the opposite. Occult forces have driven the mind of man into the darkest places of the imagination, creating morbid and obsessive interest in death and evil.

A magician consults the moon and stars.

The word occult is derived from Latin words meaning 'hide' and 'to cover over'. Where there is the occult, there will be the moon.

Astrology

Astrology is the longest-standing intellectual tradition in the world. It emerged in Mesopotamia in the second millennium BC, when it was used to predict the weather and other natural events. It quickly became linked to personal destiny. When the classical world of Greece and Rome collapsed, the ancient texts were preserved by the culture of Islam. Greek and Latin were translated into Arabic, and then back into Latin. After the tenth century AD, the Latin translations made their way into medieval Europe.

Astrological theories flourished and were taught through the Renaissance into the eighteenth century, after which they fell into

disrepute. They stopped being taught and, like alchemy, played no further part in the intellectual life of Europe. Astrology did continue as a popular pseudo-science, and as such flourishes to the present day. There have been astrological traditions in most major world cultures, including India, China and the Americas.

A sixteenth-century astrologer preparing a natal horoscope shortly after the birth of a child.

The moon's place in the Western astrological world starts in about 4000 BC with the Sumerians, who worshipped the God of the Sun (Utu), Venus (Inanna) and the moon (Nanna). Their rulers came from the priests who communicated with these gods. Banu priests emerged who could read the signs of the sky. These priests predicted natural phenomena, usually eclipses of the moon.

By 1300 BC in Mesopotamia, the precursors of birth horoscopes began to appear, and by 600 BC the twelve constellations of the zodiac were established. The Greeks assimilated mythology into astrology and this was eventually accepted and taken over by the Romans.

In the second century AD the Egyptian astronomer, mathematician and geographer Ptolemy argued that the earth was at the centre of the universe. This idea was accepted and held sway for the next 1400 years. Ptolemy also argued that astronomy and astrology were complementary. In the third book of his work the *Tetrabiblos* he argues that character is formed at the moment of conception and this character is influenced by the position of the planets. The father's influence will be governed by the sun and Saturn, and the mother's by Venus and the moon. Ptolemy's astrological work contains the seeds of modern astronomy.

The only complete manual of astrology to survive from Roman times is the *Matheseos Libri VIII* (the *Eight Books of the Theory of Astronomy*) of Julius Firmicus Maternus. A wealthy retired lawyer, Firmicus devoted the whole of Book 4 to the description of the effects of the moon on the other planets and how this would affect the planetary influences at work on the life of a man. His work had an enormous influence on the astrological thinking of the Middle Ages and the Renaissance.

THE MOON IN MODERN WESTERN ASTROLOGY

In modern Western astrology, the moon is seen as a female influence defining childhood, mother roots and learned responses. The position of the moon relative to the other signs is of great importance.

ASTROLOGY IN PRACTICE

Astrologers divide the sky into twelve houses called a zodiac. Zodiac is a Greek word meaning 'circus of animals'.

The astrologer takes the date, time and place of birth of an individual. This information is used to prepare a horoscope. The horoscope plots the position of the moon, stars and sun relative to the zodiac. The horoscope can then be used to predict events and to evaluate the course the subject's life might take. Astrologers say that the stars impel; they do not compel. They influence how and what we are but do not force us; we have and must use our free will.

A horoscope is a complex document, and the skill of the astrologer lies in the ability to interpret it. Everybody has a sun sign and a moon sign. These express where the sun or the moon was in the zodiac at the time of birth. The sun stays in each house of the zodiac for 30 days; the moon passes through every sign in a month. The sun determines personality; the moon determines emotions. The sun sign is a clue to how you will be perceived in the world. The moon sign is a clue to the voice of the inner person. The moon sign is about hopes, dreams, fears, insecurities and intuitions.

Each sun sign is linked to one of the four elements: fire, earth, air and water.

The signs of the zodiac.

The twelve sun signs: their dates and elements		
Sun sign	**Date**	**Element**
Aries	21 March–20 April	Fire
Taurus	21 April–21 May	Earth
Gemini	22 May–21 June	Air
Cancer	22 June–22 July	Water
Leo	23 July–23 August	Fire
Virgo	24 August–22 September	Earth
Libra	23 September–23 October	Air
Scorpio	24 October–22 November	Water
Sagittarius	23 November–21 December	Fire
Capricorn	22 December–20 January	Earth
Aquarius	21 January–18 February	Air
Pisces	19 February–20 March	Water

A person born in a fire sign may have the moon in a water or an air sign. This is believed to have an impact on that person's character and will have a bearing on the interpretation of that person's chart. This chart indicates the influence the moon has on character when it is in a particular sign.

General influence of the moon and the elements

Moon in:	Element	Influence of moon
Aries, Leo, Sagittarius	Fire	The moon in a fire sign will indicate a passionate, emotional person who likes to be centre stage.
Taurus, Virgo, Capricorn	Earth	The moon in an earth sign indicates a person who is unpretentious, with a no-nonsense approach to emotions. The moon will influence this person to be highly reliable and to like order and stability. They will have a tendency to be stubborn. The moon in Capricorn makes for a person who finds expressing emotions in words difficult, so prefers action.
Gemini, Libra, Aquarius	Air	The moon in an air sign indicates an outgoing, gregarious person, interested in others. The moon can cause this person to appear reserved and cold and flinch from physical contact.
Cancer, Scorpio, Pisces	Water	The moon in a water sign indicates sensitivity and a character ruled and sometimes dominated by emotions. This person is influenced to care for and nurture others, to have lots of love but possibly be over-dependent on others.

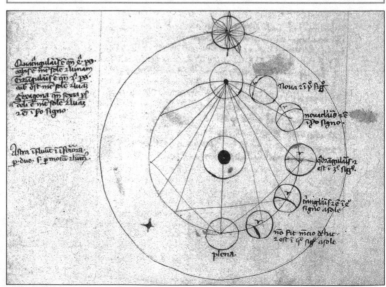

Phases of the moon from Galen's Opera Varia.

Moon phases

The phase that the moon was in at the time of birth also has an effect on personality. Establishing your birth moon phase is complicated. You need to know the longitude and latitude of your birthplace and your time and date of birth. You will also need access to a moon phase calendar.

Moon phases	Possible effect on character
New moon phase	Can cause childlike wonder and excitement about life. Brings spontaneity and enthusiasm which can lead to exhaustion. Helps make early mark in life but can bring difficulties later.
1st crescent moon	Encourages the assertive, the positive and the creative positive. Helps bring success in twenties and thirties.
1st quarter	Can make for a strong-willed, argumentative and demanding personality. A person born in this phase will be influenced to promote the new at the expense of the old. Promotes success in thirties and early forties.
Gibbous moon	May be influenced to be caring and constructive – a helper or carer. Can cause a person to give up in the face of obstacles. Success may come in middle age.
Full moon	This influence pushes a person to bring things to fruition. It tempers logic with intuition and the practical with the creative. Also causes guilty and irrational fears. This phase indicates success in late middle age.
Disseminating moon (waning gibbous)	Influences strong leadership and communications skills. Can form a revolutionary or a tyrant. Brings success in early fifties.
Last quarter	This phase can form a wise and mature person, full of counsel and understanding. A good teacher and an idealist. Tends also to encourage melancholy. Encourages success in late fifties.
Balsamic moon (last phase before new moon)	Encourages a spiritual, dreamy and contemplative personality. Intuitive and far-sighted. Understands mother nature. Enjoys a life of the mind and the spirit, rather than action and the physical. Will live through many changes. Promotes success, if at all, in later life.

Lunar auspices

Every month, the moon passes through all twelve signs of the zodiac.
It stays in each sign for about 2.5 days. As the moon changes signs,
so its influence on the subconscious changes. That influence waxes
and wanes as the moon enters, passes through and leaves each sign.
When the moon is between signs, it is said to be 'void of course'.
This is a time when things go wrong.

Effect of lunar auspices

Sign moon is in	Effect
Aries	Good time for risk-taking, decision-making and starting wars. A time when events move fast and don't last long.
Taurus	Time to start long-term plans that will be slow-moving. Not a time to change things.
Gemini	A mercurial time when things are hard to pin down. Not a good time to start things.
Cancer	Cancer is the moon's house. This is a time when people may be moody and easily upset or offended. This is also a time when people may eat and drink too much.
Leo	A time for extravagant gestures and selling ideas.
Virgo	A time for in-depth planning and dealing with details. A time to think, research, and attend to medical things.
Libra	A time for creative work, attending to friendships and romance. A time too when negative and bad emotions can be aroused.
Scorpio	A time when anger is about and suspicions are raised over financial matters. A time of criticism.
Sagittarius	A time to go away. A time when things are as they seem and are hopeful. A time for reading and pursuing ideas.
Capricorn	This is a disciplined time when the rules apply. Things are hard and frustrating during this phase. This is a time to knuckle down and get on with things.
Aquarius	A social time. A good time to present new ideas and think about the future in a practical way

Pisces	A time to see astrologers and get advice, but a time to avoid becoming confused in the hands of others. A time of introspection.
Void of course	The void-of-moon period can last from a few minutes to nearly a whole day. When this happens newly bought objects break, decisions go wrong, agreements hang, objects get lost, mistakes are made, travel is subject to delays, and accidents happen. Stick to routine matters, avoid launching new projects, signing contracts or making decisions.

Cancer: The moon's own sign

Cancer is the moon's own sign. People born under Cancer are called 'Moon Children'. People born under this sign need a safe, secure environment to retreat to. They are creative and have great emotional depth. They are in touch with their emotions but can be moody and insecure. A Cancerian can sulk and skulk away for days. The perfect partner for a Cancerian is a practical Capricorn who, after a row, will stop the Cancerian disappearing into an endless hurt silence. The worst partner for a Cancerian is another Cancer; the hurt silence after the row could go on for years.

THE DARK MOON

The Dark or Black Moon is a concept that occurs in both astrology and astronomy.

The Dark Moon in astronomy

Belief in an invisible dark moon is held on the fringes of unconventional astronomical thinking. It has two vague meanings:

▶ A hypothetical satellite, about a quarter the size of the moon, whose orbit keeps it permanently on the far side of the moon. It

Kenyon Cox's painting of Lilith.

is thought to be 750,000 miles from the earth. No claim to have seen it has ever been substantiated, and its existence is not taken seriously in the world of conventional astronomy.

▶ A description of the phenomenon by which, from time to time, in February there is either no new moon or no full moon. This will next occur in February 2014, when there will be no new moon, but the months of January and March will each have two new moons.

The Dark Moon in astrology

There are astrologers who include an earth satellite, Lilith, or Dark Moon, in their calculations. Some astrologers believe this to be a hypothetical satellite and others believe it to be real.

Lilith's presence in a natal chart has a strong and generally negative influence. It governs the dark side of a personality and encourages self-defeating and negative behaviour patterns. In a woman's chart it can indicate a strong, dominant, but sensitive personality; while in a man's it can induce despair and suicide. Lilith's presence in a chart is thought to lead to painful and obsessive relationships and great unhappiness.

THE MOON IN THE ASTROLOGY OF OTHER CIVILIZATIONS

Apart from Western astrology the two most important world astrological traditions are to be found in China and India. The moon is as important in both these traditions as it is in the West.

The moon in Chinese astrology

Chinese astrologers divide the sky into twenty-eight sections. Each section represents a day in the moon's passage through the sky, and is referred to as a lunar mansion. The position of the moon is thought to be crucial in deciding the correct actions for any given day.

The moon in Indian astrology

In Indian astrology, the moon is known as Chandra. Chandra influences the breasts, Monday, pearls, the colour white, the left eye of the male, the right eye of the female, the stomach, the oesophagus, the lymphatic system, the ovaries and the bladder. Professions related to Chandra are associated with shipping, water products, pearls, corals, agriculture and service under a woman. Chandra represents both mind and body. The moon is important in a birth chart.

A sixteenth-century alchemist at work.

Alchemy

Alchemy was a practical and a spiritual tradition. Alchemists were scientists, visionaries, dreamers, charlatans, crooks and thieves. Like astrology and astronomy, the roots of alchemy are ancient. Alchemy was practised in Europe from the tenth century until its ideas fell out of favour in the eighteenth. The basis of alchemy is distillation and purification. In the physical world, the alchemists sought to distil and purify base metal into gold. In the spiritual world, they sought to distil and purify the soul. Some alchemical practices and discoveries were genuine science and survive to this day. Alchemical writing is secret and often coded. An important alchemical idea was that knowledge and the human soul had degenerated since Adam and Eve. As a result, the body and soul had to be cleansed and purified and knowledge thrown away and relearned. The great goal was to discover the Philosopher's Stone, a substance that would not only turn base metal into gold but would provide the elixir of life and a route to immortality. Alchemical theories overlapped and were often contradictory.

Alchemists subscribed to the Greek idea of the four elements: fire, earth, air and water. They also thought that sulphur, mercury and salt had special fundamental properties which were important to the physical and the spiritual world. These three substances were called the *tria prima*.

One of the leading alchemists was Paracelsus (1493–1541). He developed a theory about the *tria prima*. He thought that sulphur represented flammability in the physical world and the soul, and emotions and desires in the spiritual. Mercury was a transforming agent in the physical world and represented the spirit, imagination and morality. Salt was solidity and incombustibility, and stood for the human body.

The moon is an essential element in the most important text relating to alchemy. That text is known as the Emerald Tablet.

THE EMERALD TABLET

The Emerald Tablet has several other names: the Smaragdien Tablet, the Tabula Smaragdina and the Secret of Hermes. It is an ancient text said to have been produced by the Egyptian moon god Thoth, who is also referred to as Hermes Trismegistus. The short and highly cryptic piece of writing claims to describe the secrets of the primordial substance and how it can be harnessed. The tablet's meaning is very obscure. It seems to state that all things come from some primal source of which the sun is the father and the moon is the mother:

> *Its father is the Sun, its mother the Moon, the wind carried it in its belly and its nurse is the earth.*

The Emerald Tablet. The oldest known version of the tablet is an Arabic translation dating from about AD 800. In the twelfth century it was translated into Latin and published as the *Secretum Secretorum*, or *The Secret of Secrets*. The Emerald Tablet was one of the most revered and important documents for European alchemists. It was central in the quest to find the Philosopher's Stone. The text has been translated by Roger Bacon, Sir Isaac Newton and many others.

The text of the tablet

Truly, without a lie, this is certain and utterly true: That which is lower is just the same as that which is higher and that which is higher is just the same as that which is lower, in order to complete the miracle of the one thing. And just as all things came about from one unity, by the thought of one

being, so all things were born from this one thing, change.
The father of this is the Sun. The mother is the Moon. The
Wind carried it in its womb. The Earth nourished it. This
is the Father of the entire world. Its goodness would remain
whole if it was spilt on the ground. You will separate earth
from fire, subtly, slowly, gently, carefully [lit: thoughtfully].
Something ascends from earth to heaven, and again descends
to earth, and receives the force of the higher and the lower. In
this way you will have the Glory of the whole world. In this
way every darkness will flee from you. This is the strength of
all strengths, since it conquers every subtlety and penetrates
every solid. Thus the world was created. From here will come
miraculous changes of which this is the method. Therefore
I am called Hermes Tripotent, having three parts of the
philosophy of the whole world. This completes what I said
about the working of the Sun.

The tablet seems to be saying that the moon is feminine and is
the supreme force known as the One Thing, the Archetypal Mother
ruling the world below. Its opposite is the sun, the Archetypal Father,
the One Mind ruling the world above. The moon, whose element
is water, is changeable and volatile and enables transformation. The
moon itself dissolves once a month and thus reflects the alchemical
process of dissolution and purification. The Philosopher's Stone
would be created by the union of the Mother Moon, the One
Thing, with the Father Sun, the One Mind. The two ideas are often
represented as the Queen Luna and the King Sun. Their offspring
would be a hermaphrodite and would represent the primal matter
from which the stone would be made. The Philosopher's Stone will
be purified in the rays of the sun and moon.

Great powers of healing and even immortality were ascribed to
the Philosopher's Stone. The stone would purify everything: metals,
the mind and the soul. Acquiring the Philosopher's Stone would be
to acquire God.

By the nineteenth century alchemy had been discredited.
Nevertheless it contributed to modern science, especially physics,
chemistry and medicine. Elements of Matter Theory can be traced to

alchemical thinking of the fourteenth century. Among the contributions made by alchemists to the development of the modern world are the invention of gunpowder, the development of ink and dyes, the production of glass and the alcoholic drink known as Aqua Vita.

Sir Isaac Newton's alchemical papers were among the most extensive in his library, and reveal that he spent a lot of time exploring the Elixir of Life and the Philosopher's Stone. The economist Maynard Keynes declared that 'Newton was not the first of the Age of Reason. He was the last of the magicians.'

Magic

The word 'magic' is very ancient and derives from the Greek adjective *magikos*, which describes the occult work of the Magians, a powerful group of astrologer priests who practised a philosophy known as Zoroastrianism.

Magicians use magic to harness a superior power to change the physical world and the lives of others. Magic has been said to have three functions: to produce, to protect, and to destroy. Magic works

Queen Luna. through spells, incantations, rites and other rituals. Magicians claim

An astrologer selling his soul to the devil.

to communicate with other beings on celestial planes and use altered states of consciousness induced by fasting, incantation, dancing, meditation and drugs.

Magic has been practised and persecuted throughout history and through all cultures. It is used with both good and evil intent. At its most enlightened, it will claim to bring peace, spirituality, calm and absolute good into the world. At its worst it is used to invoke dark forces to bring destruction and death.

The moon's position in the sky, moon gods, spirits, demons and moon-related matter can all be used to increase and focus magic forces and spell-making.

ANGELS OF THE MOON

The celestial kingdoms, of which the moon is part, are guarded by seven angels. The moon's angel is Phul, who can change anything into silver, the metal of the moon. The angel Gabriel is also thought to guard the moon, and has dominion over Monday.

THE MOON'S POSITION IN THE SKY

The phases of the moon, the day of the week and the hour of the day are all important factors when casting spells.

Phases of the moon

Generally, light spells should be done when the moon is waxing, darker spells when the moon is waning and the darkest, most dangerous spells when the moon is dark (the new moon).

Moon phases and spell-casting	
Moon phase	**Best for spells involving**
Waxing to full	**Luck, prosperity and gains** New beginnings and endeavours, planting ideas, clearing old ideas. Growing things back, accounts, money, relationships, pregnancy, promotion and health. Gaining money.
Waning	**Vengeance, discord, hatred, unhappiness and undoing** Seclusion, rest, discarding. Letting go, harvesting ideas, finishing off, selling.
Dark/new	**Death and destruction** A time to deal with enemies, seek dark justice and inflict harsh punishments.

Days of the moon

The moon's day is Monday. Medieval astrologers thought Monday was one of the luckiest days of the week. It is a day especially good for communicating with the dead and the spirit world, clairvoyance, spells for love, invisibility, discerning the truth and discovering the truth, and attaining spiritual elevation. Monday is a good day for anything connected to water, including travel and the sea. Agriculture, medicine, domestic arrangements and dreams are all well aspected if undertaken on a Monday.

In traditional Western astrology the planets which govern the days of the week are:

Sunday: Sun; Monday: Moon; Tuesday: Mars; Wednesday: Mercury; Thursday: Jupiter; Friday: Venus; Saturday: Saturn

Hours of the moon

The moon has sway over several hours in the day. The magician does not use the clock, but numbers the hours in the day and the hours in the night separately. Daylight hours are numbered from sunrise. The first hour after sunrise is governed by the planet of the day. The hours of the night are counted from sunset.

Hours influenced by the moon			
Weekday	Hours of the day, counting by moon	Hours of the night, counting by moon	Ruling planet of the day
Sunday	4th, 11th	6th	Sun
Monday	1st, 8th	3rd, 10th	Moon
Tuesday	5th, 12th	7th	Mars
Wednesday	2nd, 9th	4th, 11th	Mercury
Thursday	6th	1st, 8th	Jupiter
Friday	3rd, 10th	5th, 12th	Venus
Saturday	7th	2nd, 9th	Saturn

Some of the most influential magicians have been women, witches whose art is known as witchcraft. Hecate, the goddess of the dark of the moon, is the patroness of witchcraft.

Among the earliest recorded witches are the Witches of Thessaly in ancient Greece. They practised from the third to the first century BC. The witches of Thessalonica could 'draw down the moon', a rite in which the priestess would draw the moon's power into herself. So enabled, she could control day and night, travel across water without a ship, and fly.

Seventeenth-century witches flying with the devil.

The most famous ancient witch was Aglaonike. Aglaonike was said to be able to make the moon disappear. In reality it may have been that she could predict eclipses. It is possible that she was a skilled astronomer, or perhaps was the pupil of a skilled astronomer. The Ancient Greeks had a proverb: 'Yes as does the Moon obey Aglaonike.' She is remembered as being a sorceress.

Wicca

'Drawing down the moon' is practised by a modern occult movement known as Wicca. Wicca has no central organization or creed and takes several forms. In Wicca, the moon is revered as a threefold goddess representing the Maiden (waxing moon), the Mother (full moon) and the Crone (waning moon). Wicca rituals draw power from the moon and Wicca spells are governed by the phases of the moon. Wiccans recognize gods and goddesses. Of all their pantheon, greatest respect is given to goddesses with lunar associations. Some branches of Wicca do not recognize male deities at all.

Wicca covens are known as Groves and Esbats. They meet thirteen times a year at the full moon, when its magical power is considered to be at its greatest. Some covens also meet at the new moon, a time considered propitious for new beginnings. There is some confusion as to the difference between the dark moon, the last phase of the waning moon, and the new moon which immediately follows it and which is also dark. When 'drawing down the moon', the high priestess will use her own body in an attempt to channel the energies of the Moon Goddess. Through the agency of the priestess, the divine energies of the moon are brought into the physical world. Some Wiccan ceremonies are performed naked. Witches performing rites while naked are referred to as being 'skyclad'.

Spells

DIABOLE

Diabole was an Ancient Greek spell which invoked the power of the moon to punish enemies. The technique of the spell was to tell the moon every evil and bad thing the enemy had done, then beseech the moon to punish the perpetrator of the crimes. The magician Pachrates was said to have been so good at this spell that his employer, the emperor Hadrian, doubled his fee. On one occasion Pachrates lured a man into court, made him ill and killed him, all within the space of a day.

GRIMOIRE

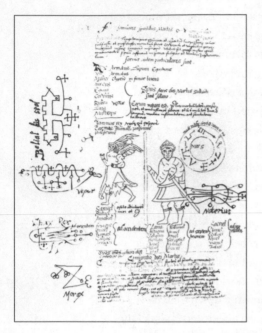

A grimoire was a textbook of spells. Many contained instructions to invoke the devil and other spirits. They appeared in the fifteenth century in Europe. The word derives from the Old French *grammaire,* which has its root in the Greek *grammatikos*, which means relating to letters. The modern words 'grammar' and 'glamour' both derive from this word. One grimoire is known as the *Grand Grimoire,* which is thought by some to have been written by Alibek the Egyptian in the sixteenth century but may actually have been written in the nineteenth century. The *Grand Grimoire* contains a very elaborate spell:

A page from a grimoire.

To cause a Girl to Seek You Out No Matter How Prudent She May Be

The spell starts with the instruction to write the girl's name on a piece of virgin parchment. After some complex instructions as to how to stand and where to look, the supplicant begins the spell with the words:

I salute and conjure you a beautiful moon, O beautiful star, O bright light which I hold in my hand! . . .

The supplicant goes on to invoke more spirits and then asks them:

To obsess, torment and harass the body, spirit and soul and five senses of the nature of N whose name is written here below in such a way that she will come to me and accomplish my will.

The supplicant puts the parchment in his left shoe and awaits results.

Simple ritual tools that can be used in lunar magic:

- An **altar**. This can be any small table that fits comfortably into your environment. It can be a rock if you are working outside, even a linen cloth spread on the ground. It must be an object on which you feel happy to put your ritual tools. The altar is the sacred centre of your world of moon magic. Choose its setting carefully, making sure that you feel only positive vibrations when you are near it.
- A **vessel** to contain liquids and moon water. This can be silver, wood, glass or horn. The vessel should be generous in proportion. The vessel will connect you to water, feminine energy, purification, healing and love.
- A small piece of **wood** you prize. This may be something you have found walking or on the seashore. It may be something you have carved or whittled. You will know it when you find it. The wood will connect you to the earth, fertility, prosperity and the home.
- A **knife** (sometimes called an **athame**). This is a ritual object. It does not have to be capable of cutting anything in real life. An ornate paper knife could do. The knife will connect you to the air, the realm of the mind and the astral.
- **Candles**. Place a black candle to the left of the altar and a white candle to the right. In witchcraft it is believed that energy passes from left to right. The black candle will absorb energy sending it to the white candle which will transmit it to the world. The candles connect you to fire, masculine energy, strength and protection.
- **Feathers**. These can be single feathers that you have collected or they can be feathers woven together and allowed to hang in the wind. The feathers connect you to the spirit, sometimes called *akasha* (Vedic) or *aether* (German).
- **Pentangle**. Placed in the middle and pointing north to define the centre of the altar. The altar should be positioned so that it and the pentangle point north.
- **Censer**. Placed in line with the pentangle. This will be used to burn incense and other aromatic substances.

THE WITCH'S BROOMSTICK

Witches have traditionally been portrayed as flying on broomsticks. These are also known as besoms. Although the besoms have acquired a sinister reputation they are really tools for sweeping away bad influences and spreading good. A simple small, even miniature, broomstick can be made from wood and dried herbs or straw. This can then be hung on a wall nearby or used as a wand.

PURIFICATION AND MOON BATHING

Before performing magic it is necessary to cleanse and purify the mind and the body. In moon magic, moon water is used.

To prepare moon water

Moon water is ordinary water purified in the light of the full moon. Place three pints of water in a silver vessel. If no silver is available, use a glass vessel with a silver coin or piece of jewellery in it. Set the vessel outside in the light of the full moon. Leave it there for several hours. The following ceremony should be performed not later than one hour before dawn. At your appointed time, kneel before the vessel, hold both hands above it, and concentrate on purifying thoughts. Imagine a white flame cleansing all around it. Imagine everything bathed in the cold white light of the moon. Mentally ask that the water should cleanse and purify you. Rise, take the water and place it in a dark cool place. Use the water to ritually anoint yourself before undertaking any moon magic or ceremony.

MOON MIRRORS

Mirrors and reflecting surfaces have always held a fascination. They have existed for thousands of years. At the beginning of the Middle Ages mirror technology improved as mercury and silver nitrate came to be used in their manufacture. This was partly thanks to the discoveries of the alchemists. Better mirrors would make the astronomer's work easier and just as the astronomer uses a mirror to reflect the moon's light and magnify it through lenses, so the magician uses the mirror to contain the moon and magnify its spiritual power.

Making a moon mirror

To make a moon mirror, start with an ordinary round mirror about 6 inches in diameter. Have to hand some fine oil and a quart of water in which mugwort has been boiled. (Mugwort is *Artemisia vulgaris*, also known as chrysanthemum weed or wild wormwood.)

Place the mirror outside, where it can directly reflect the light of the moon. If the mirror is inside, the light of the moon will be distorted or bent by refraction as it passes through the glass of the window and the process will be impaired. For the magician and the astronomer, the less glass between the mirror and the moon the better. Sit with the mirror, stare at the reflected moon and imagine the mirror absorbing the moon's energy. When the time feels right, seal the glass by smearing oil on its surface. Do this with a finger, starting at the centre of the glass and wiping outward in a spiral. Do this slowly and quietly. Then clean the mirror with the mugwort water. Use a clean linen cloth dipped in the water to wipe the oil from the surface. Store the glass in

A nineteenth-century engraving of a Saxon moon idol.

a dark place wrapped in a fine cloth such as velvet. It is important that the mirror is not exposed to sunlight.

The moon mirror can be used for:

▶ **Scrying**: seeking prophetic visions in the glass's surface. This is done at night and ideally under a full moon. It is necessary to stare into the glass, relax the mind and let the subconscious wander in the world beyond the glass.
▶ **Viewing past lives**: this is best done at night in the dark under a full moon. The magician stares into the glass and attempts to let the unconscious mind wander through the landscapes of the past.

It is important in both processes to hold the question that is to be answered very clearly in the mind. Both techniques are very difficult to perform and success is not guaranteed.

RULES FOR SPELL-MAKING

The key to successful spell-making is to have good intentions and to make careful preparations. The following are some key elements in the process of preparation.

▶ Define your goal.
▶ Define and understand your motive.
▶ Visualize your goal.
▶ Check moon phases to make sure the moon is auspiciously placed to enhance your spell-making.
▶ Write out the words you are going to use.
▶ Gather the objects you will need.
▶ Cast your circle.
▶ Invite the spirits in.
▶ Revisualize your goal.
▶ Do the physical actions required by your goal.
▶ Visualize the energy of the spell going out into the cosmos.

- Imagine it done.
- Thank and dismiss the spirits.
- Let excess energy flow from you.
- Open the circle.
- Keep a record of what you have done and why.

PREPARING THE ENVIRONMENT FOR A SPELL

Spells should not be undertaken lightly or frivolously. Just as the mind must be prepared, so must the area where the spell is to take place. Imagine yourself to be at the centre of a circle of good energy. Take your ritual knife (or a wand if you have one), move clockwise round the circle letting the tool define the circle of safe energy which protects you. Once cast, stay inside the circle if at all possible. At the end of the ritual reverse the process to release the energy back into its elements, where it may safely reside until next called upon.

Spells should not be undertaken lightly or frivolously. Just as the mind must be prepared, so must the area where the spell is to take place.

LUNAR SPELLS

It is believed that the moon can be called upon to help with many things, some of which are:

Spell for prosperity

On any night during a waxing moon, place three pieces of silver on an altar in front of you. Place a mirror on the table and in front of the mirror place a silver bowl of water.

Light the candles, saying: 'Blessed be the light of prosperity glowing and growing for mine and me.'

Grasp the silver coins and imagine them growing. Drop the first coin into the water, saying: 'All I desire is mine this night, moon and stars and candle light, money, happiness, prosperity, all these things now part of me.'

Drop in the second coin and repeat the phrase.

Drop in the third coin, saying: 'By this coin and by the light, all is mine and all is right, dark world vanish.' Stare at your eyes in the mirror.

> 'As by my will I am prosperity incarnate,
> The light of success shines on me
> So the rite is done so let it be.'

Dismiss the spirits.

Spell for better sex and increased sensual desire

Cast a circle and face south in a full or waxing moon on any day except Wednesday or Saturday. For stones use cornelian, quartz or malachite and say:

> 'May your passion gleam at thought of me
> May it heighten and grow three by three.
> When you look at me feel fire
> We will soothe that deep desire.
> Together and together we will cleave.
> So let it be.'

Spell for leaving someone

Cast a circle anticlockwise using a wand, knife and salt. Face north and cast in a waning moon or in the dark phase on a Saturday or a Sunday. Recite:

> 'I wish you well as now we part
> I follow what is in my heart
> Each must walk our separate way
> Our union over this day.'

Moon plants for use in spell-casting	
Camphor	Repel unwanted lovers and ward off colds
Cucumber	Cure headaches and enhance fertility
Gardenia	Attract lovers
Eucalyptus	General well-being and healing
Lettuce	Induce sleep and relaxation and decrease lust
Poppy	Fertility, prosperity, prophetic dreams
Sandalwood	Air purifier, healing, protection
Succulents	Love and abundance
Willow	Healing, wishes, blessings of the moon

Astral Projection

Astral projection has been practised for thousands of years and by many cultures in China, India, the Americas and Europe. The practised astral traveller leaves the body and wanders at will through time and space. The astral worlds visited might be anywhere on earth, on a higher level where the dead and non-human spirits reside, or even on a plane that is beyond time and space.

Witches and magicians have traditionally concocted substances to assist the mind into a state of altered consciousness to aid astral travel. Some recipes were highly noxious and many contained hallucinogens of one sort or another. Many of the recipes have been lost.

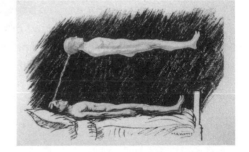

An astral projection in action. This is as far as the novice should attempt to go!

There are modern recipes that can be prepared on a domestic basis and then offered to the moon for consecration before use.

Astral travel should be attempted with the physical body in a warm, safe, protected environment. The new and full moon phases are excellent times to attempt astral travel.

A MODERN FLYING OINTMENT RECIPE

The ointment is prepared by adding the ingredients to a base of lard
or pure olive oil. A half-cup of lard or olive oil will provide enough
base for one experiment in astral projection. Mix the ingredients in
a sterilized stainless-steel or heat-proof glass container. Use an olive
wood spoon to stir the mixture. Use a safe heat such as a tea light
candle to heat the ingredients. This should all be done at night in a
place that is quiet, warm and safe. Candles are the best illumination.
A trusted companion can attend and assist. If two people make the
ointment it is important that they plan the procedure beforehand and
understand the order of events. It is also important to decide on the
'mission' – where the journey is going in time and space. The novice
is advised not to be too ambitious. To travel back three minutes
and look lovingly at your self and your partner as you prepare for
the journey is a good first journey objective. It may help to play soft
'transcendental' or quiet drumming music.

If using lard it should be softened before mixing.

Take a handful of lavender, a pinch of thyme and a half-handful
of rosemary that you have dried yourself. Grind them as finely as
you can in a pestle and mortar. Add a small handful of parsley and
grind it in the powder to form a rough paste. Pour the oil from the
sterilized container or spoon the softened lard into the mortar and
combine it with the paste. Do not rush this process. Fix your eyes on
the mortar and focus your mind on the journey ahead.

Hold the vessel containing the mixture above your head and
dedicate it to the moon. You might say out loud or chant with your
trusted companion:

> *'O Moon, I seek your help and power. On this journey keep
> me safe. Lead me and guard me. Be thou my leader, my guide
> and my friend. Endow this substance with thy great grace.
> Let the journey now be.'*

When you feel the mixture is ready (trust your instincts), smear it as
thinly as you can over your whole body. (Your trusted companion
can do this for you.) Lie quietly on the floor with your head on a

small cushion. Then focus on where you are going and drift. Your companion will stand as your physical guard as you wander.

When your journey is over sit up very slowly. Wash the ointment from your body. Take the remaining ointment and bury it where the moon will shine on it.

Then rest. Let your mind reflect on what you have done.

A CAUTIONARY NOTE ON INTENTIONS WHEN MAKING LUNAR MAGIC

When performing any spell it is wise to bear in mind the old witch's charm: 'As ye binde so are ye bound' – what goes around comes around!

Tarot

Tarot is an occult practice for divining the future using a special deck of cards. The word derives from the Italian *tarocchi*, meaning triumphs or trumps. Some occultists believe that Tarot as a form of divination dates back thousands of years to Hermes Trismegistus. It probably appeared in Europe in the fourteenth century and has been popular since the eighteenth.

The tarot reader is asked a question and uses a deck of 78 cards, each bearing a different image, to intuitively answer. The cards are laid out in patterns and the combination of the images guides the reader towards an answer. Lunar references abound in the symbolism of all the deck but two cards are directly related to the moon.

THE HIGH PRIESTESS

In modern tarot decks the high priestess is a goddess who sits with the crescent moon at her feet, and the Tora or law in her hands. She wears a crown on her head emphasizing her status. She is silent, serious and enigmatic. This is a card of wisdom and intuition, and of teaching.

The moon is shown shedding drops of dew over two towers. Two dogs bay at her and a crayfish waits, hidden in the water. This is a card of warning and of danger. Moonlight is deceptive – all is not as it seems, and the crayfish waits for those who do not heed the warning and spot the danger. The moon's influence reigns over all of nature: the air, the sky, the earth and the water; man, animals and fish. The water is dew and blood; the water of death and the water of life:

Tarot card of the moon.

> *Upon the corner of the moon*
> *There hangs a vaporous drop profound*
> *I'll catch it ere it come to ground*
> *Macbeth*, William Shakespeare

Palmistry

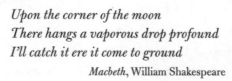

The palmist's task is to divine the past, present and future of the subject from the markings on the hand. The zodiac, the sun, the moon and the planets each have an area of the hand assigned to them, each of which governs a different aspect of career, health, emotions and relationships. The moon's area is in the mount of the moon situated in the fleshy part of the hand, low and opposite the thumb. A developed mount of the moon indicates a caring, creative and compassionate person with a wholehearted approach to life. A flat mount indicates a sense of intuition towards the feelings of others and for ideas.

Lunar Animals

Several animals have a link to the moon.

WOLF

The wolf is associated with the psychic elements of the moon. Many cultures believe that wolves howl at the moon. This has never been proved one way or the other.

HARE

Linked to fertility, the myth of the hare in the moon is found in many parts of the world: East Africa, South America and Europe. The hare is equivalent to lunar phases. The hare is also linked to the Anglo-Saxon god of fertility, Oestre, from whom the word Easter comes. The God had the head of a hare. Many Celtic, Indian, Native American and Chinese lunar deities were also depicted with the head of a hare.

FROG

The frog is a fertility symbol related to the moon. It is the bringer of rains. Hekt is an Egyptian frog goddess of fertility. The frog is also a totem animal for Native Americans of the Algonquin people. Manitou the Great Spirit lived on the moon, influencing the tides and weather of the world.

TOAD

In Chinese mythology, the three-legged toad represents the Yin, the female aspect of life. One Chinese myth has it that a lunar eclipse came because a three-legged toad had swallowed the moon. Native Americans associate the toad with the dark cycle of the moon.

CAT

The cat is associated with Artemis and Cynthia/Diana.

COW

The moon goddess worshipped in Ancient Egypt.

OWL

Linked to Hecate, the goddess of the dark moon; often seen before the death of someone close. The owl's call is a sign that the spirit of the moon is calling to help a soul back home.

Blue Moon

The phrase 'once in a blue moon', which colloquially refers to a rare event, has two rather more precise meanings, one old and one new.

The newer meaning refers to a full moon occurring twice in the same calendar month. This happens about once every two and a half years. There were about forty of them in the twentieth century. The older meaning refers to the fourth full moon in a quarter year that normally has three. In such a quarter, it is the third full moon that is the blue moon.

The phrase 'once in a blue moon', which colloquially refers to a rare event, has two rather more precise meanings, one old and one new.

Blue moons may be caused by atmospheric conditions. The material thrown up by the volcano at Krakatoa in August 1883 caused the moon to appear blue all over the world for nearly three years.

The first reference to a blue moon is found in a short treatise written in 1362 called *Rede me and be not wrothe*:

Yf they seye the moone be belewe
We must beleve that it is trewe.

The blue moon is said to have a face that talks to everything in its light. A blue moon is usually taken as unlucky, although it can also be seen as a time to sow new ideas and ambitions that will take a long time to grow. Care must be taken when releasing dreams and hopes under a blue moon, as their potency will be very strong.

A blue moon is usually taken as unlucky, although it can also be seen as a time to sow new ideas and ambitions that will take a long time to grow.

The child born under a blue moon has great potential but it may never be realized unless the child grows into an adult that can harness the power given to it by the double moon. The mature blue moon child will be compassionate, caring and sensitive; the immature adult will be volatile, moody, emotional and even cruel.

In the season of a blue moon, the tides will be high and the weather stormy.

RECENT AND FUTURE BLUE MOONS

19 August 2005: Third full moon in a season of four full moons
30 June 2007: Second full moon in a calendar month
21 February 2008: Third full moon in a season of four full moons
31 December 2009: Second full moon in a calendar month
21 November 2010: Third full moon in a season of four full moons
31 August 2012: Second full moon in a calendar month
21 August 2013: Third full moon in a season of four full moons
31 July 2015: Second full moon in a calendar month

The Moon and Numerology

The moon's number is 2 and its magic number is 9.

Moonstone

Moonstone is a semi-precious stone linked to the moon and said to bring luck. It is an opaque feldspar (of which the moon is actually partly made – see the chapter on 'Facts and Figures'). It exists in many different colours, including silvery grey, milky white, green white, pale pink and brown, or pale yellow. It looks like dew from the moon or tears, and in India is known as 'the tears of the moon'. The history of the moonstone is rich in myth and superstition.

Ancient Greeks and Romans wore amulets of moonstone to ward against madness and epilepsy, and hung it in orchards and gardens to promote plant growth. Also called selenite, moonstone is associated with the goddesses Diana, Selene and Aphrodite. The Romans believed the crystal was created by the light of the moon. It was used in ceremonies for Diana, whose image they believed was contained within the stone.

In the East, moonstone is one of the luckiest stones. The strange way it reflects light implies that some beneficial spirit is inside.

In the East, moonstone is one of the luckiest stones. The strange way it reflects light implies that some beneficial spirit is inside. In India it must traditionally be displayed for sale on a sacred yellow cloth. You must stare into the stone during a waxing moon to divine the future. It arouses tender passions and is a traditional gift for newlyweds and new lovers. Moonstones are a birth stone for the month of June. Lovers should hold the stone in the mouth during a full moon to divine the future. The finest examples are called adularia or rainbow moonstone and come from Sri Lanka.

The Moonstone Temple at Anuradhapura in Sri Lanka, built in 100 BC, was reputed to have had steps which were inlaid with an incredible mosaic of moonstones. The steps have long since been looted.

Superstitions

Births

Traditional	Baby born during a full moon is a child of fortune.
Traditional	Hold a newborn baby up to the light of a new moon and give thanks and prayers for a long life.
Lithuania	Wean boys on a waxing moon; wean girls on a moon that is waning.
Cornwall	A child born between an old moon and a new moon will never see sixteen.
Traditional	No moon, no man. (A child born when the moon is in a dark phase will die before he becomes a man.)
Cornwall	When a boy is born on the wane the next child will be a girl.

Domestic

Traditional	If the moon shines on washing left out at night, it shines on the clothes for a funeral.
Traditional	Move house on a new moon.
Devonshire	Cut hair and nails in a waning moon.

Destiny

Popular, traditional	Before you go to bed, place a prayer book under your pillow. It should be open at the wedding service, at the point where it says: 'with this ring I thee wed'. Place on it a key, a ring, a flower and a sprig of willow, a small heart cake, a crust of bread and the nine of clubs, nine of hearts, ace of spades and ace of diamonds, all wrapped in a thin gauze or muslin handkerchief. Get into bed, cross your hands and say: 'Luna, every woman's friend, Let thy goodness condescend; Let me this night in visions see Emblems of my destiny.' If you dream of storms you will have trouble; if the storms end in calm so will you. If you dream of a ring or the ace of spades marriage is likely.

≫

A dream of bread means an industrious life.
A dream of cake means a prosperous life.
A dream of a willow tree means treachery in love.
A dream of spades means death is close by.
A dream of diamonds, money, clubs, a foreign land or keys means wealth will come to you.
A dream of birds means you will have many children.
A dream of geese means that you will marry only once.

Luck (good)

Devonshire	See new moon over right shoulder, lucky; over left, unlucky; and straight ahead is good fortune until the end of the moon.
English	On the first evening of a new moon, women should sit astride a gate and say: 'A fine moon, God bless her.' This will bring good luck.
Arabian nomads	Eat mashed peas on which the new moon has shone. This will bring luck.
Traditional	A new moon on Monday is a sign of fair weather and good luck.
Iran and Egypt	On seeing the new moon, quickly close your eyes and open them to stare on your beloved. This will bring luck.

Luck (bad)

Traditional	Sleep with the moon on your face and you will have nightmares or become mad.
England and Scotland	'A Saturday moon If it comes once in seven years Comes once too soon.'
Staffordshire	Seeing the new moon for the first time through trees is unlucky.
Traditional (for women)	Do not let the moon shine on your face while sleeping – it will twist your face and rob you of your beauty.
Traditional	Pointing at the moon brings bad luck; pointing at the first new moon of the year brings twelve months' sorrow.
Norfolk proverb	'Saturday new and Sunday full Never was good and never wull.'
Northern Italy	Change of moon on a Wednesday is dreaded.

Medical

Traditional	To cure warts, wait until moon is twenty days old, then lie on footpath to receive its light and gaze upon it. Rub everything within reach.
Old French, traditional	Rub warts with dirt while looking at the moon – this will cure them.
Traditional cure for child with a cough	Take the child out and let it look at the new moon. Rub your hands over its stomach, saying : 'What I see may it increase, What I feel may it decrease In the name of the Father, Son and Holy Ghost, Amen.'
India	Bathe naked in the waters of a river and face the new moon to overcome infertility.

Romance

Traditional	Unlucky to marry during the waning quarters of the moon or in May. Lucky to marry during waxing quarters and June. Do not let the moon shine upon your nuptial bed: it will bring bad luck.
Traditional	At the first new moon of the year, look at the moon through a new silk handkerchief that has never been washed. The number of moons you see through the handkerchief will be the number of years you will be unmarried.
Traditional	On the first new moon of the new year, take your stockings off and run across a field. When you reach the other side of the field you will find between the great toe and the next a hair, which will be the colour of your lover's.
Traditional	On the first new moon after midsummer, young people should go to a stile and say: 'All hail new moon, all hail to thee! I prithee, good moon, reveal to me This night who shall my true love be; Who is he (she) and what he (she) wears, And what he (she) does all months and years.
Yorkshire	At the first new moon of the new year stare into a looking glass for the first sight of the new moon to reveal how many years before you will be married. The number of years depends on the number of reflected moons.

»

Ireland	After saying, 'In the name of the Father, the Son and the Holy Ghost,' point at the moon with a knife and say: 'New moon, true morrow, be true now to me That I ere the morrow my true love may see.' The knife must then be placed under a pillow and strict silence observed until dawn or the spell will be broken.
Berkshire	At the new moon, girls must go into a field and say: 'New moon, new moon, I hail to thee! By all the virtues of my body. Grant this night that I may see He who my true love shall be.' After this they go home to sleep and their future husband will appear in their dream.
Traditional	Marry during a full moon.
Traditional	Become engaged within two days of the full moon.

Wealth

	At a new moon
Traditional	Touch silver in your pocket and say: 'You find us in peace and prosperity so leave us in grace and mercy.'
Traditional	Turn money in pocket and think about being lucky.
Traditional	Shake your pockets, take out all the money and let the rays of the new moon shine on them.
Traditional	Count your money in the light of a new moon – it will increase.
Traditional	Stand on soft ground under a new moon, turn your money over, make a wish and turn round three times.
	On first day of the first new moon of the year
Traditional	Shut your eyes and turn the smallest silver coin in your pocket upside down for luck and prosperity all the year.
	If you don't have silver coins
Traditional	Turn your apron over in the light of the new moon to bring a month of luck.
Traditional	Bow three times to the new moon and make a wish.

Traditional	Turn your apron over three times and a present will arrive before the next new moon; or curtsey three times and say: 'Welcome, New Moon, I hope you bring me a present very soon.'
Traditional	Make a wish and it will be realized before the end of the year.
Traditional	If you live in a country ruled by a king visit him when the new moon is one day old and ask for what you want. He will give it to you.
Traditional	When you see the new moon, kiss the first person of the opposite sex you see without speaking and you will receive a gift.
Ireland	On seeing the new moon, kneel down and say the Lord's Prayer. Then say: 'May thou leave us as safe as thou hast found us.' Cross yourself and say: 'In the name of the Father, the Son and the Holy Ghost, Amen.' And then wish for whatever you want.

Weather and farming

England	'If the Moon change on Sunday there will be a flood before the month is out.'
Traditional	'An old moon in a mist Is worth gold in a kist [chest] But a new moon's mist Will never lack thirst.'
Traditional	Set eggs under a hen at new moon.
Traditional	Root up trees when the moon is on the wane but after midday.
Suffolk	Don't kill a pig on a waning moon lest the pork be waste in the boiling.

Chapter 7

Medicine, Madness, Werewolves and Science

Over 30,000 years ago, one of our ancestors carved what we think are the phases of the moon on to a piece of bone. That ancestor's work was an early attempt to understand what the moon is doing. In the millennia that have passed since the carving was made we have used two broad methods to understand the moon and the heavens. One method is physical, the other psychological or spiritual.

The physical method is based on observation and written records. After centuries of observation it became possible to say what the moon was going to do next because it was known that it had done it before. This knowledge carried with it supernatural overtones. The person who could predict something as frightening as an eclipse was surely more than just a scribe with a measuring stick; surely that person was in touch with the unseen forces that bound the world. Such knowledge held power. And at that point the psychological or spiritual method comes into play. The observer takes advantage of the power he or she wields. The discoverer of what is happening becomes the guardian of the sacred secrets, a priest who is in touch with the gods and to whom is given the knowledge of why things are happening. A priest can play on the fears, hopes and dreads of the uninitiated, can promise great rewards to the virtuous and terrible punishments for the wicked; a priest has power. Science, magic, alchemy and prophecy mingled and, in the West, stayed mingled for nearly 3,000 years until the beginnings of the Enlightenment in the seventeenth century.

The peoples of Mesopotamia who first recorded the motions of the moon on to clay tablets in a writing known as cuneiform were astrologers, but their methods were those of the astronomer.

The records made in Mesopotamia for the purposes of prediction contained meticulous scientific data collected with as much precision and accuracy as their instruments allowed. By 474 BC a cycle of lunar eclipses had been detected that repeated every 19 years. The Greek thinker Meton's discovery, in 432 BC, of the 19-year cycle that now bears his name is thought to have been based on the work of Babylonian scribes.

The mixed scientific and mystical method of thinking pervaded every area of thought that we now define as scientific. Astrology, with its links to the cosmic and heavenly forces and its power to predict, was a very important influence.

This was especially true of medicine, a discipline that deals with matters of life and death and by extension punishment and forgiveness. A doctor operating in ancient Greece might be well-versed in the contemporary theories about the nature of illness and have a lifetime's clinical experience with patients. Nevertheless his prognosis as to a patient's survival, or his decision to operate or bleed, would take into account the astrological implications of the positions of the moon, sun and stars.

Astrology is the oldest intellectual discipline in the world. It was taught and treated with great respect until the middle of the seventeenth century. In the second century AD the Greek doctor Galen asserted that the complete physician must be well-versed in astrology, and 1500 years later the herbalist and physician Nicholas Culpeper said much the same thing, asserting that 'physic without astrology is like a lamp without oil'.

Ideas about the moon that were formulated in the classical world made their way through the centuries to medieval Europe and emerged as a mixture of alchemy, astrology, science and superstition. At the end of the Middle Ages the power of the ancient Greek and Roman theories declined. Copernicus placed the sun at the centre

of the universe and Ptolemy was banished from the new heliocentric *Claudius Galen.* world. These were the first glimmerings of objective scientific thought. Astrology gradually lost its footing as a science. By the time Darwin wrote *The Origin of Species* in 1859, the idea that astrology and medicine were linked was no longer taken seriously by scientists.

But astrology did not vanish. The nineteenth century saw a rebirth of interest in the occult. With industrialization the city rather than the field became the habitat for most people. Gas and electricity gave power over light itself. Night and day, the seasons and the so-called natural rhythms of life were obscured. The industrialized mind turned to superstition and the occult. Dracula, Frankenstein's monster, Jack the Ripper, Dr Jekyll and Mr Hyde were among the lead players in the phantasmagoria of terrifying creatures that lurked in the imaginary dark corners of the new urban world. Moonlight caused fear and death, it brought madness and horror.

Societies sprang up that practised magic and pagan rites. These movements provided a popular new home for astrology, a home where the moon could hang on to her supernatural powers.

There were some scientists who could not let go of the idea that the moon held physical sway over man's mind and body. There is still a vigorous debate on the subject, clouded by popular myth and superstition. To date, no broad theory that empirically links the phases of the moon with human behaviour has emerged.

The Moon and Medicine in Greece and Rome

Ancient classical medical theories started with the idea of man as a microcosm of the great macrocosm, the universe. The Greeks saw patterns repeated at all levels of the cosmos. The individual was not only a microcosm of the human race but of everything. The individual contained patterns that are in harmony with everything in the universe. Illness occurs when that harmony is disturbed, and the doctor's job is to restore harmony. Alchemy was central to this idea and is summed up in the alchemical precept: 'As above, so below, as without, so within.'

Greek and Roman doctors believed that:

- The sun, the moon, the planets and the stars exerted an influence over everything, including plants and stones.
- Each part of the body was ruled by a different sign of the zodiac.
- The planets influenced the openings of the body.
- The planets had influence over treatments and medicines.

- 'Critical Days' were of great importance. These were related directly to the moon.
- The moon influenced acute fever.
- The sun influenced chronic fever.

The Greeks and Romans inherited from the Egyptians a developed system of astrological mathematics applied to medicine called *Iatromathematica*. An element of this system established a detailed relationship between the cosmos, stones and plants. From this it was deduced that the moon had a tangible power over all earthly things, and especially over anything that was connected with water. Four thinkers about science and medicine who put great store by the moon stand out: Hippocrates, Pliny, Ptolemy and Galen.

HIPPOCRATES
460–370 BC

Hippocrates is a very important figure in the development of medicine and is sometimes referred to as 'the father of medicine'. The Hippocratic oath traditionally taken by physicians to affirm the sound ethical basis of their work is attributed to Hippocrates or one of his pupils. He rejected the idea that the supernatural or divine caused illness. At the base of his medical theory was a belief that illness was caused by an imbalance in the body of the four humours or fluids: blood, black bile, yellow bile and phlegm.

Four Humours.

The four humours

	Temper	Organ	Nature	Element
Black bile	Melancholic	Spleen	Cold dry	Earth
Phlegm	Phlegmatic	Lungs	Cold wet	Water
Blood	Sanguine	Heart	Warm wet	Air
Yellow bile	Choleric	Bladder	Warm dry	Fire

Hippocrates thought that the moon had power over the whole human body. An important element in Hippocratic medicine was the idea of 'Critical Days'. These days were days counted after the first day of an illness and linked to the moon. If a patient had a crisis on one of these days it would indicate a relapse or even death.

Hippocrates thought that the moon had power over the whole human body.

Hippocrates is said to have written that a physician without knowledge of astrology had rather call himself a fool than a physician.

PLINY THE ELDER
AD 23–79

Pliny the Elder was a Roman who wrote a *Natural History* (*Historia Naturalis*), in which he describes the influence of the moon on farming and on living creatures. He wrote:

> *We may certainly conjecture, that the moon is not unjustly regarded as the star of our life, this it is that replenishes the earth. When she approaches it she fills all bodies, while when she recedes she empties them. For this cause the blood of man is increased or diminished in proportion to the quality of her light.*

PTOLEMY
AD 85–165

Ptolemy was a polymath who counted astronomy and astrology among the many disciplines of which he was master. Ptolemy wrote the *Almagest* (*The Great Work*), which attempted to sum up all knowledge; *The Geography*, which attempted to describe the known world; and a work on astrology, *Tetrabiblos* (*The Four Books*). In the *Tetrabiblos* he is believed to have used astronomical data collected hundreds of years earlier by the Babylonians in Mesopotamia. Ptolemy is the most influential of the ancient thinkers. He argued that the earth was at the centre of the universe and his ideas held sway for 1500 years. For Ptolemy, astrology was something that could be used in all parts of life and especially medicine. He taught that the planets influenced the four medical elements of heating, cooling, moistening and drying (fire, earth, water and air). He believed that astrology was an important factor and must be taken into account when making a medical diagnosis. Other factors might be the race, age and upbringing of the patient. Ptolemy had great faith in the work of the Egyptian astrologers.

GALEN
AD 129–c.200

Galen took on and developed the ideas of Hippocrates. Like Ptolemy, his teaching was very influential. His ideas survived to dominate medical theory into the Middle Ages. The moon played an important part in his theories. He too believed that the position of the moon relative to the astrological houses at the time of illness was a clear indicator of whether you would live or die. Galen wrote a treatise called *Prognostics from the Time of the Patient's Taking to His Bed*. In this he stated:

> *Whoever will know the condition of the patient at the beginning of when he is ill, how long he will endure it, let him look at the ascendant and the moon . . . [and if] the Moon*

conjoins with the benefices, and then if the patient became ill
at that hour and it is the beginning of his illness, it indicates
recovery from his illness . . . if the patient is released as long
as the moon stays in a place and a term away from death, as,
when it reaches the benefices or when they aspect it, his pain is
lightened and he is released from his agony.

Galen's *Prognostics* is an extension of Hippocrates' Critical Days. The phases of the moon were used to calculate the Critical Days: 7, 9, 14, 18, 21 and 28 days after a new moon were all important moments in the development and cure of illness. If the stars aligned with the moon to produce good influences on those days, then all would be well with the patient. If the aspects were bad, then so it would be with the ill person. Galen constantly refers to the moon in his writing and like Ptolemy he states that he is following the Egyptian astrologers. He was confident that their work and observations about the moon were true.

Galen's 129 books on medicine were translated into Arabic in the ninth century and then into Latin in the eleventh century. Galen's medical writing became the basis of much of the medical thinking in the Middle Ages.

The Sphere of Democritus

Astrological and medical theories became very complicated and systems came into being that would help physicians interpret the influence of the moon and stars. One system is called 'The Sphere of Democritus', which consists of a rectangular diagram with the days of the month arranged in three colours. The physician had to find out under which day of the moon the sick man took to his bed. He would then perform an elaborate calculation which took into account the position of the moon on the day the patient fell ill, a number derived from the patient's name and the day of the month, and a division of the whole of this by thirty. The resulting number would then be referred to on the 'sphere', and according to where it fell the patient would live or die. If it fell in the lower part he was doomed. If it fell in the upper part, all would be well.

Democritus.

The aspects of the moon were also an important factor when considering the best time to take medicine. A treatise by Dorotheus of Sidon written in the first century BC states:

> *If you commence drinking the medicine for diarrhoea, then the commencement of it is best when the moon is in Libra.*

The Moon and Medieval Medicine

Although the great empires of Greece and Rome faded and vanished, their ideas did not. They were preserved and refined by Arab scholars and by the eleventh century AD were finding their way into medieval Europe.

The next two centuries also saw the founding of some of the great European universities in Bologna, Padua, Paris, Montpellier and Oxford. Astrology continued to be a dominating influence.

Medieval astrologers held that the cosmos, the stars, the planets, the moon and the sun influenced growth, decay and the seasons and therefore played an important part in agriculture and medicine.

The enormous silvery object in the night sky dominated and confused the astrologers. Sometimes it was thought to be bigger than the earth; sometimes, to have beams that could penetrate the infested atmosphere of the earth. The astrologer scientists were certain the moon had special strength and powers, and the writing of the ancient Greeks and Romans seemed to confirm these ideas. So they must be true.

Medieval physicians believed that:

▶ Illness was a punishment from God.
▶ The human body was a microcosm of the universe.
▶ The body was made up of the four elements, earth, air, fire and water.
▶ The elements had to be kept in harmony for a being to be healthy.
▶ The moon was of the greatest importance in preserving that harmony.

Like the rest of science, practical medicine was a combination of prayer, astrology, spells, mysticism and treatment based on records and observation.

One of the most important scholars of the early Middle Ages was St Isidore of Seville (560–636). He wrote the *Etymologiae*, an encyclopaedia of all knowledge. It ran to twenty volumes and had an enormous influence throughout the Middle Ages – it could be found in every monastery library. St Isidore's influence lasted into the Renaissance. Between 1470 and 1530, ten editions of the *Etymologiae* were printed. In the *Etymologiae*, he advocated medical astrology and the notion that man was a microcosm of the universe. At the centre of his theories on medicine was a certainty about the importance of the moon.

The following examples give some idea of the chaos that reigned for hundreds of years as religious dogma and the occult fought with sound medical thinking and insights.

There is a document known as the *Anglo-Saxon Herbal* which contains prescriptions and advice for the cure and treatment of illness. Many of the entries refer to the moon. In a cure for lunacy using cloverwort:

For a lunatic take this wort and wreath it with a red thread and weave it about the man's neck when the moon is on the wane.

The medieval physician shared the Greek idea that the moon had to be taken into account not just in serious operations and blood letting but also in the preparation of medicines themselves. In one example among hundreds the *Anglo-Saxon Herbal* advises that:

Periwinkle must be plucked when the moon is 9 nights old or 11 or 13 or 30 nights.

Medicine and predictive astrology were sometimes interchangeable. An eleventh-century manuscript holds that:

On the 4th day of the moon let blood in the morning. This is a good day on which to begin undertakings and send boys to school. Whoever runs away will speedily be recaptured. Whoever falls ill will quickly die with no hope of recovery. Boys born on this day will be fornicators and so will girls. If you have dreams good or bad on this day it will come true.

Another eleventh-century text, *On the Make-Up of the Celestial and Terrestrial Universe*, has in it a description of the consequences of the birth of a woman when the moon is in Capricorn and Aquarius:

Whatever females that are born by day with that horoscope will be masculine women, quick to take up farming, waging war and building houses and neglecting the scents, sandals and clothing belonging to women. They seek that which is common to men, nor are they effectively opposed in this by their husbands.

There were glimmerings of a crude psychoanalytical theory linked to the moon. Monastic libraries contained *Lunaria de Somnis*, books that catalogued the influence of the moon on dreams. One claimed that to dream of the moon meant joy, and that a dream made on the sixteenth or seventeenth day of the moon would come true.

All over Europe, cults for the moon goddess Diana grew up. Around the cult was woven a complex web of superstition. According to one source:

> *Wicked women who have given themselves to Satan and have been seduced by phantasms and illusions of demons believe they can ride with Diana Pan Goddess and a huge throng of women on chosen beasts in the hours of the night. They say that in the silence of the night they can travel great stretches of territory, that they obey Diana as though she were their mistress and that on certain nights she calls them to her special service.*

Equating inherited truths with the evidence of observation and practical experience was as difficult in the Middle Ages as it is today. Men and women who might win Nobel prizes in the modern world had their thinking mangled and twisted by ignorance and dogma. Traditional beliefs wrapped rational thought in an impenetrable fog.

An interesting case is the Abbess Hildegard von Bingen (1098–1179). She has been described as a polymath. She was a musician, visionary, linguist and physician. Unusually for a woman of the period she travelled widely and communicated with popes and emperors. She wrote a series of notes about medicine, possibly for use in her monastery, St Rupertsberg, at Bingen on the Rhine. She too held the Greek medical theories of the four elements and the idea that illness came from an imbalance within the body of these elements. Among her writings she talks about the influence of the moon on the human body and in the preparation of medicines:

> *The moon in phases has influence on the fluid balance of human beings. Now the phases of the moon do not rule human nature as if she were a goddess nor can human beings apply anything to or withdraw anything from the moon . . . blood and other humours can be moved according to lunar phases.*
>
> *A strong man should be bled every three months because of the two changes of the moon the blood has its maximum strength.*

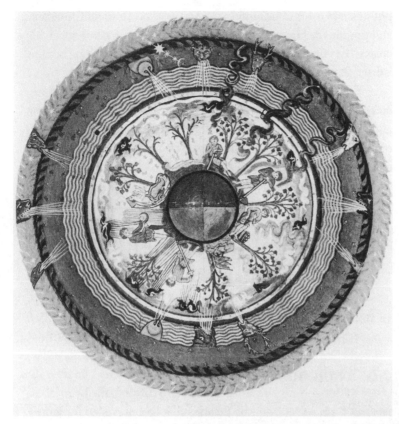

The moon and stars can be seen at the top of this illustration from Hildegard von Bingen's 'Celestial Influences on Men, Animals and Plants'.

*If noble and healthy herbs are picked under a waxing moon
then they will prove better for preparation of medical pastes.*

Hildegard von Bingen had a developed idea of the influence of the
moon on sex and procreation:

*When the blood increases in the organism with the waxing
of the moon, the human being man or woman is ready
for procreation for when the moon is waxing and there is
increasing supply of blood the seed is strong.*

By the fourteenth century the system of astrological medicine had
become very complicated, too complicated in fact for the ordinary

physician to use it. The moon was used as a way to simplify diagnosis and prognosis. The moon was the fixed point in the astrological charts, the one thing that could be pinned down. Everything became subservient to the position of the moon relative to the zodiac.

Finally, in the sixteenth century some sort of order began to emerge. As new scepticism questioned traditional ways of thinking and Copernicus, Galileo and Kepler placed the sun at the centre of the universe, inherited ideas, superstitions and religious dogma about the nature of the universe gave way to rational thought. The Enlightenment had begun.

The Enlightenment and Lunar Medicine

Claudius Galen's Recetario de Galieno, 1518.

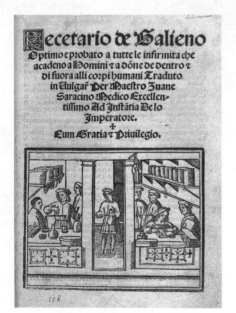

In 1628, William Harvey published *Exercitatio Anatomica de Motu Cordis et Sanguinis in Animalibus* (*An Anatomical Exercise on the Motion of the Heart and Blood in Animals*). In it he shot an arrow straight through the heart of the theories that had prevailed for more than two millennia. He described with complete accuracy the method by which the heart pumped blood round the body. For centuries physicians had followed the teachings of Galen. Galen had taught that there were two types of blood. They were produced by the heart and the liver, and the blood so produced ebbed and flowed like tides to different parts of the body. Harvey's ideas were not immediately accepted but it was the beginning of the end for the old medicine. At the same time, ideas that all things were made up of earth, air, fire and water were replaced by a growing understanding of new disciplines, including chemistry, physics, biology and anatomy.

Bloodletting without anaesthetic.

The new order required proof based on observation. It demanded laws that would accurately predict what was happening in the universe. Those laws would be the result of an empirical exploration of the rational world. Experiments would be devised that could be repeated to establish facts, and from those facts the theories would be deduced.

But there was a problem. Contrary to the beliefs of the rationalists of the Enlightenment, scientific thought is not just a question of laying one proven fact on another to arrive at a theory of everything. Theories are vulnerable to error, misconception and vanity. Scientists, like the rest of us, can be blinded by prejudice.

Isaac Newton was a central figure in the Enlightenment, and has been described as its father. He has also been described as the last of the alchemists. His published works are models of rational thought. He described universal gravitation and framed the laws of motion. His work on mechanics is the basis of modern engineering. In 2005, the Royal Society said that Newton was more important than Einstein. It is ironic that the great mind that grappled with motion and gravity also struggled with alchemy and magic.

Newton had rationally and elegantly used the microcosm of a falling apple and the macrocosm of the moon to deduce that the same laws of motion and gravity that governed the apple also governed the orbit of the moon round the earth. The same rational mind believed that the union of the archetypal Father Sun with the archetypal Mother Moon would produce a hermaphroditic representation of primal matter from which would be conjured the Philosopher's Stone. Newton wasted untold hours looking for this chimera. Books and papers on alchemy formed a major part of Newton's library.

Even as the Enlightenment waxed, the moon waned as a central power in medicine, but its influence did not go away. While the body can be dissected and its mechanics analysed, this is not true of the mind. The mind became the last prisoner of the moon's power.

Eighteenth-Century Definition of a Lunatic

In the mid eighteenth century Sir William Blackstone (1723–80) wrote his *Commentaries on the Laws of England*. The commentaries were a bestseller. One of the commentaries deals with the treatment of the insane, describing a lunatic as:

> *One who has had understanding but by reason of disease or grief has lost the use of his reason. A lunatic is indeed one that has lucid intervals, sometimes enjoying his senses and sometimes not and that frequently depends on the change of the Moon.*

This definition was used in the framing of the Lunacy Act of 1845.

Those deemed to be 'lunatics' would be monitored and restrained with chains or other devices at times when it was thought

A seventeenth-century Dutch engraving of moonstruck women. The men on the left are their deserted husbands.

they would be violent and a danger to themselves or others. New and full moons were high on the list of critical times.

The idea that the moon had a measurable effect on the human mind floated around the edges of scientific thinking throughout the nineteenth and into the twentieth century. It was fed by the apparent truth that natural disasters and deranged human behaviour happen most often when the moon is new or full.

It is true that there is a wealth of anecdotal evidence that hospitals, the police and fire services are all at their most busy during a full moon. Similarly disasters, floods, accidents and other dramatic events do happen around a new or full moon. But it is very difficult to find any evidence that this is inevitable rather than coincidental. Attempts to use science to examine and classify this influence have proved confusing and contradictory. It has not proved easy to apply the rules that control the physical world of gravity, light, electricity and magnetism. The objectives and methods of the investigators have been confused and contradictory.

However, in the 1960s a new and apparently coherent theory about the moon and mental health emerged. It is known as Cosmobiology. Its main proponent is Dr Arnold Lieber. His theories are contentious, and Lieber has legions of detractors as well as supporters. His work is held by some to be at best bad science and at worst crackpot. His theories rely heavily on the interpretation of statistical evidence.

Cosmobiology: the Biological Tide Theory of Human Behaviour

Cosmobiology is the most developed and comprehensive theory about the influence of the moon on the human mind and body to emerge since ancient Greece. The theory considers the body to be a microcosm of the universe mirroring the effects of the forces at work in the cosmos. It states that the human frame is subject to the same gravitational, atomic and electromagnetic forces as the universe. It is due to imbalances in the body caused by those forces that the human being suffers from illness and disturbed mental behaviour. The theory argues that the moon in particular exerts an observable gravitational influence over water, and that as the body is made up of 80 per cent water, the moon must have a similar pull on it. This theory has been around in one form or another for thousands of years.

The theory goes beyond water. The argument continues with an assertion that the moon's power combines with other forces in the universe to cause fluctuations in the earth's atmosphere and electromagnetic field. This happens on a periodic basis and is called a biological tide.

It then asserts that mankind is especially susceptible to biological tides because we are covered in skin. Skin is a semi-permeable membrane that allows the movement of electromagnetic forces in both directions. The body is continually trying to hold a balance between gravitational, electromagnetic and nuclear forces. These forces ebb and flow like tides within the body, especially the parts of the body that contain or secrete water (just as these forces cause the tides to ebb and flow on the earth's surface). This influence is exaggerated by the fact that the body's nervous system contains sensors that are sensitive to fluctuations in the earth's electromagnetic field.

The pineal gland is used as an example. The pineal gland receives gravitational and electromagnetic information and reacts to produce melatonin and serotonin, which, the theory claims, impacts on sexual and reproductive rhythms. (In fact the exact function of melatonin in the human body is not clear. Reduced levels of melatonin are thought to be a contributory factor in causing cancer.) The pineal gland is situated deep within the brain. In humans the gland is important for sexual development and shrinks after puberty. It should be understood at this point that the pineal gland has attracted many myths, superstitions and metaphysical theories as to its function. It has been described as the 'fifth chakra' (a chakra is held to be a 'force centre' in the body). It is claimed to be a dormant organ, which, if activated, will enable telepathy. It has also been claimed that it secretes the hallucinogenic chemical dimethyltryptamine (DMT). The exact function of DMT in the human body is not clear.

According to Cosmobiology, the biological tides are most disturbed when gravitational forces are at their strongest, which is at the time of the full and new moons. These are therefore the times in which there will be an increased chance of natural disaster and when the body itself is most disturbed. That is why at full and new moons human beings are liable to present with a series of symptoms, which range from headaches, bloatedness and bad temper to full-blown attacks of psychotic violence. In 1982 Dr Lieber put his ideas into practice. He argued that an alignment of the moon and planets was going to cause a major earthquake in California. It didn't.

SOME OF THE ARGUMENTS AGAINST THE 'LUNAR EFFECT'

Some of the main arguments against the 'lunar effect' and Cosmobiology are:

▶ The raw data is analysed in ways which are confused and which tend to corroborate the theory.
▶ Phenomena related to the lunar effect have not proved to be predictable or repeatable.
▶ Many of the assumptions are based on misconceptions, e.g. the moon has a strong effect on unbounded water and as the human

body is 80 per cent water the moon has a strong effect on the human body. In fact the water in the body is not like the oceans unbounded, it is a closed, bounded system.

▶ Proponents of the lunar effect use unsubstantiated special pleading. Claims such as: 'It is obvious that . . .'; 'The evidence states . . .'; 'We all know that . . .'; 'Work has been done which proves . . .' are used without any evidence to support the claim being made.

▶ Supporting evidence is often anecdotal and subject to common reinforcement based on self-deception and delusion. For example, hospital staff will agree that the A&E department is most busy during a full moon because others have reported that this is the case even though they may not even be aware of what phase the moon is in.

▶ Some of the evidence derives from mythology reinforced by collective hysteria and superstition. This is a version of the modern urban myth. For example, there is a legend that vampires operate in this area. There is a report of vampirism in the next valley. X says that he knows someone from the next valley who saw a vampire. Therefore the myth must be true so we will bolt our doors at night and hang garlic in our rooms.

The two camps for and against the lunar effect theory are very entrenched.

Modern Astrological Medicine and the Moon in the Twenty-First Century

Astrology is still used in the practice of medicine. Today's astrophysician would find a lot in common with his or her medieval counterpart. Like the ancient Greeks, they take a holistic view of illness. The universe, the body and the soul are linked and illness happens when there is an imbalance between those three things.

Some even hold that illness is the result of a sin or misdemeanour performed in a past life, the paying of a karmic debt. A case study cites a man who had been blinded and scarred in an explosion. The man displayed symptoms of psychological disturbance. Part of his treatment was to explain to him that he was paying for cruelties he had performed in ancient Egypt.

In contemporary astrological medicine the moon is as important as it ever was. The body is considered to have several 'chronobiological rhythms' that are indicators of our sleep patterns, our blood pressure and hormones. The moon influences them all.

In astrological medicine the moon is seen as being one of the most important factors in how a child grows up, imprinting patterns of thought and behaviour that stay ingrained for the rest of its life. The moon controls something as general as our habits and as particular as the left eye of a man. It is the moon that will alter our moods, change our blood pH and our uric acid levels. Operations are trickier in a new moon and easier when the moon is void (i.e. is between one astrological house and another).

In astrological medicine the moon is seen as being one of the most important factors in how a child grows up

The twenty-first-century astrological physician would have a lot to talk about with his or her medieval or ancient Greek counterpart. The work of seventeenth-century astrological physician William Lilley forms part of the theory and practice of modern astro-medicine. For the astrological doctor, things haven't changed much for 2000 years.

The Moon, Menstruation and Childbirth

It is often asserted that the moon regulates the menstrual cycle and that childbirth increases during full moons. It is very hard to prove either claim. The same statistical evidence that is used to prove the claims can be used to disprove them. Different scientists have taken the same hospital records of childbirth and come up with contradictory conclusions. There are similar problems with the statistical analysis of suicide and accident records.

Menstruation varies from woman to woman and can be different from month to month for the same woman. The average menstrual cycle is 28 days but that is as far as it goes. The cycle in some women can be as low as 20 days, or as high as 40 for others. Only 30 per cent of women have a menstrual cycle that matches the average. Other mammals have cycles that range from 5 days for a rat, 37 days for a chimpanzee and 14–16 weeks for an elephant. None of them appears to have any lunar link.

Misconceptions about the power of the moon are buried very deep in the human mind. They emerge as myth, and myth can quickly turn into 'fact'.

Lycanthropy and Werewolves

There is an ancient belief that at a full moon certain individuals will change into animals. The technical name for this is lycanthropy.

Lycanthropy has two meanings, one medical and the other mythological. The word itself comes from the Greek *lykanthropos*, which means 'wolf man' *(lykos* = wolf; *anthropos* = man). The full moon has come to be associated with both the medical and the mythological meanings. The mythology may have contributed to the symptoms of the medical condition.

MEDICAL MEANING OF LYCANTHROPY

Clinical lycanthropy is the delusion on the part of the affected person that they have been changed or are in the process of being changed into a wolf or other animal. There are records of the other animal being a hyena, a cat, a horse, a tiger, a frog or even a bee. There is one case of serial lycanthropy where the patient thought he had progressed from being a dog to a horse to a cat.

It is not known what causes lycanthropy. It is possible that it arises in part from a medical condition such as schizophrenia,

A werewolf attacking children.

bipolarity or clinical depression and that the delusions are made worse by inherited cultural traditions to do with werewolves and vampires. Sufferers have been shown to have unusual activity in the parts of the brain that are to do with the recognition of body shape.

MYTHOLOGICAL LYCANTHROPY

In the myth a person turns or is turned into a wolf or other animal. The change is often associated with a phase of the moon, especially the new or full moon. The myth has many forms and is found in Europe, Africa, South and North America and the Far East. In some cultures the transformation is seen as being a spiritual gift and the person able to perform it may be revered as a priest or shaman.

ORIGIN OF THE LYCANTHROPY MYTH IN THE WEST

The original myth has several variants. They all deal with King Lycaos, a monarch with many sons, one of whom was called Nyctimus. Lycaos kills Nyctimus and attempts to serve him on a plate to the god Zeus. Zeus is disgusted and pushes the plate away. He then turns Lycaos and his remaining sons into ravening wolves.

In another version Lycaos sacrifices Nyctimus to Zeus, and immediately the sacrifice is transformed into a wolf.

Both Zeus and Lycaos are associated with light. Nyctimus is associated with darkness. Deep behind this myth may be the rituals of a cannibalistic tribe who recognized the wolf as their totem. The slaying of Nyctimus by Lycaos is a representation of the triumph of light over dark, day over night, the sun over the moon.

PETRONIUS: *THE SATYRICON* AND WEREWOLVES

In *The Satyricon*, Petronius tells a story of lycanthropy that is directly linked to the full moon. The story tells the fate of the slave Niceros who travelled by the light of the full moon to meet his mistress, an attractive widow. Niceros is accompanied by a soldier who is a guest of his master. They arrive at a crossroads which is surrounded by tombs. The soldier begins to behave strangely. He takes off all his clothes, piles them in a heap and makes a circle of urine round them. He then turns into a wolf and disappears, howling into the night. His clothes are turned to stone.

The terrified Niceros carries on to meet his mistress and when he arrives she tells him that her house has been visited by a wolf and that the wolf has killed several sheep. A servant has chased off the wolf by stabbing it in the neck with a spear.

The next day Niceros returns home. At the crossroads he cannot find the heap of stone clothes but the ground where they were last seen is soaked in blood. He reaches home to find the soldier, now restored to human form, lying in bed. He is being tended by a doctor and has a bad wound in his throat.

In France there is a strong tradition of werewolves. In 1693 in Benais a wolf-like creature is reported to have killed over 100 victims. Between 1764 and 1767 a creature known as the Beast of Gévaudan is reputed to have attacked up to 210 people, killing 113, many of them children. Between 1809 and 1813 in Vivarais it is claimed a wolf-like beast devoured twenty-one victims.

Gilles Garnier

Gilles Garnier was a man of sombre countenance. He had a long grey beard and his eyebrows met across his head. He lived in Dole in France.

One night, 8 November 1573, the night of a new moon, a small group of peasants passed close by a wood. Suddenly they heard a girl screaming. They also heard the howling of a wolf. They rushed into the wood, where they found a badly injured girl and saw a sinister shape bounding away into the darkness.

Attacks on children and adults continued in the area. Permission was granted by the authorities to hunt the werewolf and 'to tie and kill it without incurring any penalties'. Garnier was caught apparently in the middle of an attack on a child. He was tortured and confessed to being in the power of a force outside himself and admitted to taking the flesh of his victims home to his wife for her to enjoy. Garnier was found guilty and sentenced to be burned. Many people swore that they had seen him in his wolf form. It is very likely that Garnier was simply the victim of collective hysteria.

Peter Stubb, the Werewolf of Bedburg

In 1589 Peter Stubb (also known as Stube, Stubbe, Stumpf) was arrested and charged with lycanthropy at the age of forty. After torture on the rack he confessed to having practised the 'dark art' since the age of twelve. He admitted that the Devil had given him a magic leather belt and that when he wore it he turned into a wolf. He claimed that as a wolf he had eaten cattle, children and pregnant women. He also claimed that after killing the pregnant women, he tore out the foetuses and ate them. He also confessed to incest with his daughter.

Stubb was sentenced to a terrible death. The flesh was removed from ten parts of his body with red-hot pincers. His arms and legs were broken with a wooden axe. Horribly maimed, he was beheaded and burned. His daughter and mistress were convicted as accomplices and they too were burned. The magic belt was never found, and the charges were not proved.

MODERN REPORTS OF WEREWOLVES

23 March 1933, Zimbabwe

A young Scottish doctor reported secretly observing a ritualized jackal dance. The male dancers ate rotten meat, drank an intoxicating potion and had sex with the surrounding women as though they were jackals on heat. He claimed to have seen two of the dancers transformed into jackals.

Daily Telegraph,15 July 1949

The paper carried a report of a werewolf in Rome:

> *Howls coming from the bushes in the garden in the centre of Rome last night brought the police patrol to what seemed to be a 'werewolf'. Under the full moon they found a young man, Pasqual Rissini, covered in mud, digging in the ground with his fingernails and howling. On being taken to hospital Rissini said that for three years he had regularly lost consciousness at periods of the full moon and found himself wandering the streets at night, driven by uncontrollable instincts. He was sent to a clinic for observation.*

Mr W., 1975

Mr W. was a 37-year-old Appalachian farmer who had served in the US Navy. Shortly after his discharge he began to show symptoms of lycanthropy. He lost interest in his job, let his facial hair grow, slept in cemeteries and would lie down in the road in front of approaching traffic. Mr W. would howl at the moon and said that he had become a werewolf. Doctors were not able to say anything more than that he had an unspecified brain syndrome that presented as chronic schizophrenia. They were unable to explain why his psychotic phases occurred during a full moon.

Mr H., 1970s

Mr H. had taken LSD while serving with the US Army in Europe and had felt himself turning into a werewolf. His hallucinations included being able to see fur growing on his face and hands, and he said he could look into the Devil's world. He also suffered from an overwhelming desire to chase rabbits. Mr H. was admitted to hospital and on examination it became clear that he had a long history of taking hallucinogenic drugs. He suffered from paranoid delusions, delusions of grandeur and acute schizophrenia. After medication he stopped hallucinating and the delusions vanished, though he continued to show a morbid interest in Satanism.

POSSIBLE NON-LUNAR CAUSE OF LYCANTHROPY

Porphyria

Congenital porphyria, a very rare disease, is a possible cause of the werewolf legend. The disease is caused by the over-production of porphyrins and manifests itself in mental and physical disorders. Porphyria takes two forms: acute porphyria, which is inherited; and cutaneous porphyria. The British royal family has a history of acute porphyria, the most famous sufferer being George III.

The symptoms include the ulceration of skin lesions, progressive mutilation of the physical structures such as ears, nose, and eyelids, and grotesque distortion of the fingers. The skin can become pigmented and the teeth may be stained red or reddish brown by the deposits of porphyrins. Victims may have abnormal hair growth, especially across the forehead. Urine may change colour to red when exposed to sunlight. The victim may become sensitive to bright light and seek the dark. Mental disorders associated with porphyria include hallucinations, depression, anxiety and paranoia.

Given the right cultural circumstances these symptoms could be misread as evidence of werewolfism. The fearful, superstitious rural communities of Europe in the Middle Ages provided just such circumstances. A person suffering from the disease, found wandering through the moonlight with distorted features and abnormal hair growth, his teeth and nails stained red, with open sores on his body, could easily be misinterpreted as a werewolf fresh from a bloody encounter, his skin torn by the fight and his teeth and nails stained with the blood of his gory acts.

Life Forms That Are Known to Respond to the Moon

Some life forms, though, can be shown to respond to the moon, especially to moonlight.

The Palolo worm
Palola viridis
The Palolo worm's reproduction cycle is linked to the phases of the moon. It shares this with other marine invertebrates, which include corals, urchins, sea cucumbers and crinoids.

The Palolo worm lives in burrows. It reproduces at fixed times of the year and in cycles which synchronize with lunar phases. As spawning time approaches, the rear end of the worm develops paddles and eyes. At a moment which is triggered by moonlight, the worm backs out of its burrow. The newly developed rear part breaks away and swims to the surface. It appears on the surface in the early morning during the last quarter of the moon. Twenty-eight days later, during the next last quarter of the moon, the cycle is repeated and the worm reappears in even greater numbers. In the Pacific, the worms reproduce in October and November. In the Atlantic the sequence takes place in June and July. In the laboratory the worms have been shown to be sensitive to moonlight. The behaviour of the worms can be changed by manipulating or switching off artificial moonlight. The worms are extremely sensitive to light levels and can detect light invisible to the human eye. They are aware of the sun when it is below the horizon even when it is obscured by muddy water. The worms also respond to tidal and solar stimuli, which regulate other biological clocks within them.

The reproductive sections of the worm are considered a great delicacy.

> *The Palolo worm lives in burrows. It reproduces at fixed times of the year and in cycles which synchronize with lunar phases.*

Flatworm
Convoluta roscoffensis
This worm lives on the seashore and appears and reappears in anticipation of the tides. When taken into the laboratory far from the sea it will continue to repeat the tidal cycles for several days.

Sea urchins
Echinoidea
The mature form of this creature spawns at the full moon.

The banner-tailed kangaroo rat
Dipodomys spectabilis
During the winter, the banner-tailed kangaroo rat is nocturnal. On bright moonlit nights it will not leave its burrow.

The larvae of the antlion
Myrmeleon
These will excavate large traps during the full moon.

MIGRATION AND THE MOON

Many animals use the tides or moonlight to orientate themselves. Some animals use the motion of the tides to help them move long distances to and from spawning grounds. The North Sea Flat Fish journeys to its spawning grounds by rising into the tide when it is going in the right direction and sinking to the bottom when the tide would push it in the reverse direction. Many animals use a version of this technique.

The Sooty Tern
Onychoprion fuscatus
The Sooty Tern has a complex migratory pattern that seems to be linked to a lunar cycle. The bird lives on Ascension Island near the equator in the mid-Atlantic. After dispersing over the Atlantic, the birds return to the island around the time of every tenth full moon. They lay their eggs about two full moons later.

The amphipod crustacean
Talitrus saltator
This is a type of crab that lives in the sand of the Mediterranean at the high-water mark. It makes nightly forays to the land and returns to its burrowing zone just before dawn. If these animals are placed in glass containers in the laboratory and they can see the moon, they will gather towards where the sea would be. If the moon is invisible they show no directional preference. If the moon is reflected with a mirror and made to appear to come from a different direction, the crabs will generally cluster towards the 'new' direction of the sea.

The sea slug
Tritonia diomedea
The slug seems to use the earth's geo-magnetic field and the full moon to orientate itself. At the new moon when there is no light, the slug's directional movements are random. At full moon, it will tend to move in the direction of the earth's magnetic field. If, during the full moon, the magnetic field is artificially reversed, the slug's orientation is similarly reversed.

Oysters in Connecticut

Oysters in a laboratory in Connecticut were observed to open and close in a cycle that corresponded to the tides in the sea from which they had been taken.

Cockroaches, stress and the moon

Experiments with cockroaches in America have indicated that stress-related changes in the chemical composition of their blood are linked to the new and full moon. It is speculated that this is related to changes in the earth's electromagnetic field, caused by the phases of the moon.

With all these examples there are other factors at work. It is possible, though, that moonlight plays some part in an animal's sense of direction.

The Moon and the Creation of Life

The latest theory about the creation of the moon describes a collision between the earth and a Mars-sized object. On impact, some of the metallic content of the colliding body joined the earth's core and some of the shattered mantle spun away to form the moon. This happened about 4.5 billion years ago, when the earth itself was in the process of forming. At that time the earth would have had a very hot atmosphere, which would have prevented the formation of organic life. The collision would have done two things.

▶ It would have stripped the earth of its hot primordial atmosphere and made it easier for living organisms to come into being.
▶ After the impact, the earth's core would have heated up over a long period and produced a very strong magnetic field (which the earth does have). The magnetic field would have shielded the molecules of emerging life from the destructive forces of cosmic rays.

The moon is unusual in that it is very large and exerts a strong gravitational pull on the earth. The moon's gravity has stabilized the earths 'obliquity' (the amount that it wobbles as it spins) and has caused the earth's seas to be tidal. These two factors may also have had a powerful influence on the evolution of life on earth.

TIDES, OBLIQUITY AND THE EVOLUTION OF LIFE

Obliquity – the earth's wobble factor

The earth does not wobble very much and that is why its climate is comparatively stable. The planet's potential to wobble is heavily damped by the gravitational pull of the moon. Even small changes in obliquity cause life-threatening climate changes. The last ice ages may have been caused by a minor shift in the earth's obliquity. Large changes would cause chaos. Once the earth had acquired a moon, it stopped wobbling and developed a climate that was stable enough to allow the formation of organic life.

The tides

Life in its most primitive form began to appear on the earth about 35,000 million years ago. This was about 700 million years after the formation of the earth's crust. At that time the earth's surface was mainly water. The moon created tides, which pulled the water to uncover and cover islands of the mantle. The water would have been a thin soup, made up of the basic ingredients for the formation of organic life.

To form organic matter the components would have to combine and react in a process known as polymerization. Pre-cellular structures have been created in the laboratory by creating a cycle of wetting, drying and heating. This is what might have happened 35,000 million years ago. The beat of the tides revealed the newly formed earth crust and left a residue of pre-organic material that was then heated by the sun. When the tides came back in, the substances were washed back into the ocean. The cycles would repeat twice daily for hundreds of millions of years. As time went on the material in the oceans became more concentrated and eventually began to perform long sequences of chemical reactions, until a self-replicating organism was formed.

If the theory is correct then the existence of a large close moon with a planet may be a useful indication of the possibility of life on that planet.

What Does the Moon Do?

There are many misconceptions about what the moon does to the earth and its inhabitants.

Contrary to popular belief, suicides do not peak at the full moon in midwinter. They peak in high summer, and there is no conclusive evidence that they are linked to the moon at all. The menstrual cycle of a woman may average out at 28 days but this does not prove they are subject to lunar influence.

The moon does cause our seas to be tidal. All day, every day, billions of gallons of water slosh across the earth's surface, entirely controlled by the moon. Does the moon do the same things to the water in our bodies? The moon's power over the water in our bodies has been calculated at 1 part in 30 trillion; its effect over our blood as 3 parts in a million of our weight or 0.01 of an ounce in a 200-pound man. These proportions are too small to count.

Perhaps, though, size doesn't matter. Sharks can detect blood in water at levels as low as 1 part in a million. Animals can sense earthquakes sometimes days before they happen. Cattle have been observed to habitually graze with their bodies aligned to the earth's magnetic field. A hot needlepoint is nothing compared to the mass of our bodies, but it can cause excruciating pain. These are all examples of tiny forces which have a big impact. The same could be true for the power of the moon.

The moon does cause our seas to be tidal. All day, every day, billions of gallons of water slosh across the earth's surface, entirely controlled by the moon.

The power of the moon may be tiny in relation to the human frame – it has been compared to a mosquito sitting on the shoulder. Its measurable influences on our bodies may be minute but its power could be enormous. Here in the twenty-first century the jury is out. Nobody knows for certain what the moon can and can't do to us.

Chapter 8

Miscellany

Fascinating Facts about the Moon in Our Lives and Culture

Moon Idioms

Mooning	To expose one's buttocks
Blue moon	To expose one's buttocks on a cold day
Blue mooning	To expose one's genitals instead of buttocks
Moon	To behave or move in a dreamy or listless manner, usually associated with being in love
To be over the moon	To be very happy
Moonlighting	Holding down more than one job at once
Moonshine	Non-existent rubbish; or illicitly distilled or smuggled alcohol
Moonstruck	Lost in fantasy
Mooncalf	A foolish person
Moonraker	A native of the county of Wiltshire, taken from a traditional story about men trying to rake the moon out of a pond, where they could see its reflection
Once in a blue moon	Referring to something that rarely if ever happens
Moonwalk	Dance popularized by Michael Jackson, supposed to characterize the weightlessness of walking on the moon
Moonie	A member of the Unification Church, named from its founder, Sun Myung Moon
Moonstomp	Informal dance, characterized by heavy stamping
Honeymoon	Holiday taken by newly married couple. Originates from the sixteenth century, referring to affection waning like the moon after the first month

Food and Drink

- **Blue Moon:** a bright blue ice cream popular in the mid-west US. It has been described by the *Chicago Tribune* as a 'Smurf-blue marshmallow sweet, and tasting remarkably like fruit loops'.
- **Blue Moon beer:** a Belgian white beer manufactured by Coors.
- **Blue Moon cocktail:** a cocktail based around blue curaçao. There are many variations, but the most popular are made with gin or vodka. A recipe for a Blue Moon cocktail:

> 2 parts gin
> 1 part blue curaçao
> ice
> twist of lemon peel

Mix the gin and the blue curacao. Pour over ice, add a twist of lemon peel, and serve.

- **MoonPie:** a type of cake invented in about 1917. A MoonPie consists of two round crackers with a marshmallow filling, dipped in chocolate. A mini-MoonPie has been developed that is about half the size of the original. The pies come in several flavours, including chocolate, vanilla, strawberry, banana, lemon and orange.

Apollo astronauts who are Masons	
Buzz Aldrin	Apollo 11
Gus Grissom	Apollo 1
James Irwin	Apollo 15

Hoaxes

THE GREAT MOON HOAX

On 25 August 1835 the *New York Sun* ran the first of six articles describing life on the moon. The discoveries were attributed to the leading British astronomer Sir John Herschel. The article claimed

A VIEW OF
THE INHABITANTS OF THE MOON,
AS SEEN THROUGH THE TELESCOPE OF SIR JOHN HERSCHEL.

Top: A view of the inhabitants of the moon from the New York Sun, *1835. Bottom: Sir John Herschel.*

that he had used an 'immense telescope' to detect trees, oceans, beaches, goats, unicorns and other fantastic creatures on the moon. The article was written under the name of Dr Andrew Grant, who claimed to be Herschel's travelling companion and amanuensis. Grant did not in fact exist. It is possible the real author was satirizing the Reverend Thomas Dick, 'The Christian Philosopher', who had worked out that there were more than 21 trillion inhabitants of the solar system, over 4 billion of which lived on the moon. The headline to the story read: **Great Astronomical Discoveries Lately Made By Sir John Herschel, L.L.D.F.R.S.&c**

One assertion in the piece was that there might be gold on the moon:

> *We frequently saw long lines of some yellow metal hanging*
> *from the crevices of the horizontal strata. We of course*
> *concluded that this was virgin gold.*

At first Herschel was amused, but to his annoyance he was dogged
for years by people asking questions about these 'discoveries'.

EDGAR ALLAN POE'S MOON HOAX

In June 1835, in the *Southern Literary Messenger*, Poe published
'The Unparalleled Adventure of One Hans Pfaal', a supposedly true
story. The story went that Pfaal had travelled to the moon using a
revolutionary new balloon and a device for turning a vacuum into
oxygen. The journey took nineteen days. He had lived on the moon for
five years and sent a message back to earth using a local moon inhabitant.

THE APOLLO MOON LANDING HOAX

Conspiracy theories are everywhere. The holocaust didn't happen,
Elvis is alive, the CIA blew up the World Trade Center, British
Intelligence murdered Lady Di. It must be true: all you have to do is
read the evidence, play the record backwards or watch the film. There
will always be anomalies. The unexplained explosion, the fourth shot,
the white car. The moon is no exception. Conspiracy theories have
been around since Apollo 8 did (or did not) orbit the moon:

- NASA went to the moon but is unwilling to tell us how they really
  got there or what they found.
- NASA went to the moon but did not land a man there. Manned
  landing records have been faked.
- Nobody went to the moon and the whole thing is a fake.

The most developed hoax theory claims that the Apollo programme was constructed in the Nevada desert using the expertise of Disney film studios, the science-fiction writer Arthur C. Clarke and the director Stanley Kubrick.

Evidence for and against the moon-landing conspiracy

Anomalous shadows in lunar photographs

Pros: Shadows in lunar photographs appear to contradict the position of the sun.
Cons: The lunar surface is highly reflective and uneven. The cameras were fitted with wide-angle lenses. Both these factors would lead to distortion and anomalies in photographs.

Visibility of stars in lunar photographs

A footprint on the moon taken on 20 July 1969.

Pros: Photographs taken on the moon show no stars.
Cons: Cameras exposing for the brightness of the moon's surface would not be able to register stars in the lunar sky.

Astronaut footprints too well preserved

Pros: Moon dust is very fine and would not hold imprints with the clarity seen on the photographs.
Cons: The dust is the lunar regolith and it exists in a vacuum. This causes the dust particles to stick together. Astronauts described the sensation of walking on the regolith as being like walking on talcum powder or wet sand.

Presence of dust in the lunar atmosphere

Pros: The downthrust from the Lunar Excursion Modules' engines and the movements of the astronauts would throw up dust that would hang in the air. The lunar photographs show no such dust.

Cons: The moon has very low gravity and its atmosphere is too thin to cause particles within it to decelerate. In such conditions, dust particles and debris thrown up by the astronauts and their equipment would quickly fall back to the surface in a parabola like a stone thrown on earth.

Photographic backgrounds

Pros: Backgrounds in photographs taken very far apart appear to be identical.

Cons: The moon's surface seen at eye level has very few defining characteristics. Large background features are a long way away. It is very difficult to estimate distance on the moon. (The crew of Apollo 14 at one point did not know where they were or that they were standing on the edge of the crater that was their destination.) This accounts for backgrounds which are very similar but not identical.

Flares and hotspots in photographs

Pros: Flares and hotspots in lunar photographs suggest film lighting equipment placed out of shot.

Cons: The lunar regolith contains glass droplets created by volcanic acitivity. The regolith is in any case very reflective. This caused flares and reflections in the photographs. Similar flares and hot spots are created by the high-visibility jackets worn by police and emergency crews seen in newsreel footage.

Positioning of cross hair registration marks in lunar photographs

Pros: Camera registration marks used for photographic reference should appear in front of the image. In some lunar photographs they appear behind objects or misplaced.

Cons: Problems of overexposure created by the very high-contrast ratios on the moon's surface would cause bleeding between bright and dark objects on the film negative. This would account for the apparently anomalous positioning of cross hairs. Photographs for media distribution are subject to editing by NASA, creating anomalous cross hair positioning or even the disappearance of cross hairs altogether.

Flags planted by Apollo crews move as if blown by the wind

Pros: In some sections of some of the footage taken on the moon the American flag can be seen to move.
Cons: The flag appears to wave because of the methods used to tether it and because the astronauts are moving it around, causing undulation of the flag's fabric in the friction-free, airless atmosphere. Other fragments of footage show a completely inert flag.

Apollo moon landing.

The Lunar Excursion Module landing marks

Pros: The 17-ton LEM left marks which were not as clear as the footprints left by the much lighter astronauts.
Cons: The astronauts' feet were much smaller than the landing pads on the module. They exerted a greater pressure on the lunar surface and left clearer prints. For comparison a stiletto heel will make a much deeper mark on wood than a running shoe worn on the same surface.

Camera film and the lunar surface

Pros: The lunar surface is so hot that the film in the cameras would have melted.
Cons: The film was never in direct sunlight so could not overheat. The moon has no atmosphere to conduct heat from its surface up to the camera equipment. Heat could be transmitted by radiation and this was prevented by suitable protective coatings on all vulnerable surfaces, including the film stock.

Solar radiation in the Van Allen belt

Pros: The astronauts would have been killed by radiation from the Van Allen belt and from the ambient radiation in space.
Cons: The astronauts spent about 30 minutes crossing the Van Allen belt. The radiation they received was the equivalent of a chest X-ray. However, a high proportion of the astronauts who went to the moon developed cataracts. This could have been caused by exposure to cosmic rays and would be evidence to prove that they did travel through deep space.

Large telescopes and the visibility of the landing sites

Pros: Large telescopes capable of seeing tiny objects millions of light years away cannot see the lunar landing sites.
Cons: To see the flag left by Apollo 11 requires an earth-based telescope with a 200-metre-wide lens. The largest telescope on earth has a 10-metre lens. The Hubble's can only see objects 60 metres across.

Death of Apollo 1 astronauts

Pros: The Apollo 1 astronauts were murdered because they were going to reveal the NASA conspiracy to fake the moon landings.
Cons: They murdered President Kennedy too. The astronauts did not die. They were kidnapped and taken to a distant planet far away to live with Buddy Holly.

Hoaxers-in-chief

- Deke Slayton, NASA chief astronaut in 1968, visited Stanley Kubrick on the set of *2001: A Space Odyssey*. He referred to the making of the film as being at 'NASA East'.
- Stanley Kubrick is claimed to have produced most of the footage for Apollos 11 and 12.
- Frederick Ordway was involved with the Apollo programmes and worked for Stanley Kubrick on *2001*.
- Harry Lange was also involved with the Apollo programmes and worked for Stanley Kubrick on *2001*.
- NASA: the greatest hoaxers of them all.

Reactions to the Moon Race

Man's Reach

Cartoon by Herblock celebrating man reaching the moon.

Our passionate preoccupation with the sky, the stars and a god somewhere in outer space is a homing impulse. We are drawn back to where we came from.

Eric Hoffer

Prometheus is reaching out for the stars with an empty grin on his face.

Arthur Koestler

The immemorial moon – the Moon of the Myths, the Poets, the lovers – will have been taken from us forever. Part of our mind, a huge mass of our emotional wealth, will have gone. Artemis Diana, the silver planet, belonged in that fashion to all humanity: he who first reaches it steals something from us all.

C. S. Lewis

Treading the soil of the moon . . . palpating its pebbles, tasting the panic and splendour of the event, feeling in the pit of one's stomach the separation from Terra . . . there form the most romantic sensations an explorer has ever known.

Vladimir Nabokov

The Lunar Society of Birmingham

The Lunar Society was an informal group of eighteenth-century amateur scientists. Members met in each other's houses on the Monday nearest the full moon. The bright moon illuminated the way home and gave its name to the club. The Lunar Society had a very important influence on the development of modern science.

Leading members of the Lunar Society:

Mathew Boulton (1728–1809)   Joseph Priestley (1733–1804)
Erasmus Darwin (1731–1802)   William Small (1734–75)
Thomas Day (1748–89)   James Watt (1736–1819)
Richard Lovell Edgeworth (1744–1817)   John Whitehurst (1713–88)
Samuel Galton (1753–1832)   William Withering (1741–99)
James Keir (1735–1820)

*Meeting of the
Lunar Society.*

For a definitive account of the Lunar Society, you couldn't do better than read Jenny Uglow's wonderful book *The Lunar Men: Five Friends Whose Curiosity Changed the World* (Faber, 2003).

Medical

- **Moon blindness**: Periodic opthalmia. An affliction of horses that affects one or both eyes. It was named 'moon blindness' in the 1600s, when people believed that it was linked to the phases of the moon. Disappears and recurs intermittently. Usual prognosis is ocular failure and permanent blindness.
- **Moonface**: A symptom of Cushing's Disease (exogenous hyperadrenocorticism). A large round face.
- In Ayurvedic medicine, the moon is associated with **circulatory disorders**. To combat these, Ayurveda recommends cucumber.
- **Lunacy**: A term that is now obsolete, referring to insanity. These mental health disorders, characterized by periods of lucidity and insanity, were thought to be influenced by the phases of the moon.

The Plant and Animal Kingdoms

Blue Moon Butterfly (*Hypolimnas bolina*)	A species of nymphalid butterfly
Moonrat	Rat-like mammal of the hedgehog family
Blue Moon	A variety of lilac rose
Moonflower (*Ipomoea alba*)	An American tropical climbing plant with sweet-smelling white flowers, which open at dusk and close at midday
Moon Daisy (*Leucanthemum vulgare*)	Also known as Ox-Eye Daisy and Dog Daisy
Moonwort	Any fern of the genus *Botrychium*, especially *Botrychium lunaria*; so called because of its crescent-shaped frond segments
Moonseed	Climbing plant of the genus *Menispermum*
Moonfish	• US marine fish (*Vomer Setipennis*), also known as a Blunt-Nosed Shiner, Horse Fish, and Sun Fish • A thin silver marine fish (*Selene vomer*), also known as a Lookdown and Silver Moonfish
Mooneye	Any species of the American freshwater fish of the genus *Hiodon*
Moonie	Alternative name for the Goldcrest (*Regulus regulus*)

Some Moon Films

▶ *La Lune à un mètre* (*The Astronomer's Dream*), 1898: A 4-minute
hand-coloured short film by George Méliès. An astronomer is
tormented by the Devil, who makes him see a gigantic moon. After
seeing fairies he is eaten by the moon. In the last seconds Phoebe,
the goddess of the moon, saves him and gives him back his life.

▶ *Le Voyage dans la lune* (*Voyage to the Moon*), George Méliès, 1902:
14 minutes long. Fairies beat up creatures from the moon.

▶ *When the Man in the Moon Seeks a Wife*, Percy Slow, 1908: A
silent black-and-white film. The man in the moon arrives on earth
by hot-air balloon, looking for a wife.

▶ *Le Clair de lune espagnol* (*The Man in the Moon*), 1909: A
short silent film about a man travelling by balloon to the moon,
featuring anthropomorphic moon and stars.

▶ *Moonstruck*, 1909: A silent French film which describes a
sleeping drunk fighting off lunar inhabitants.

▶ *The Woman in the Moon*, 1929: German silent film about love
and exploration of the moon. The director was Fritz Lang, who
had an affair with his leading lady. It was the highest-grossing film
of the 1929/30 season.

▶ *Boom in the Moon*, 1946: Buster Keaton is mistaken for a modern-
day Bluebeard and is conned into piloting a moon launch.

- *Cat-Women of the Moon*, 1953: Gold is found on the moon and the crew are chased back to earth by beautiful cat-women.
- *Nude on the Moon*, 1960: Explorers discover a colony of topless aliens, mostly women.
- *Moon Pilot*, 1962: An explorer meets a space cowgirl who shows him how to coat his spaceship to avoid going moon mad.
- *The First Men in the Moon*, 1964: Space explorers land on the moon only to discover that the British got there in 1899. The British crew is attacked by a giant caterpillar that eats a female crew member. They escape and lunar civilization is destroyed by the common cold.
- *Those Fantastic Flying Fools*, 1967: A circus midget is launched to the moon, but only makes it as far as Russia. Loosely based on the Jules Verne novel *From the Earth to the Moon*.
- *2001: A Space Odyssey*, 1968: Classic Stanley Kubrick film, based on the novel by Arthur C. Clarke.
- *A Grand Day Out*, 1994: Wallace and Gromit build a spaceship and fly to the moon, only to discover that it is made of cheese. Possibly Wensleydale.

Classical Music with Moon Themes or References

Moonlight Sonata (1801), Beethoven
Au Clair de Lune (*By the Light of the Moon*), Debussy
La Terrasse des Audiences du Clair de Lune (*The Terrace of Moonlit Audiences*), Debussy
'Song to the Moon' appears in Dvořák's opera *Rusalka* (1900)
'Everyone Feels the Joy of Love', from *The Magic Flute*, Mozart
To the Moon, Schubert
Pierrot Lunaire, Schoenberg
To the Moonlight, MacDowell
Borobudur in Moonlight, *Java Suite*, Godowsky
Aries Moon (1938), Ruff
Aria of the Moon, Petr Cvikl

Popular Music

Bad Moon Rising (John Fogerty, 1969), by Creedence Clearwater Revival

Bark at the Moon (Ozzy Osbourne, 1983)

Be Careful When They Offer You the Moon (Pete Atkin–Clive James, 1970)

Blue Moon (Richard Rodgers–Lorenz Hart, 1934)

Blue Moon, song by German synthpop band De/Vision

(*Blue Moon* has also been the title of country albums by Toby Keith [1996] and Steve Holy [2000], a blues album by Robben Ford [2002] and an album by Swedish musician Sofia Talvik [2005])

Blue Moon of Kentucky (Bill Monroe, 1946), originally by Bill Monroe, subsequently by Elvis Presley, Patsy Cline, Ronnie Hawkins, Rory Gallagher, LeAnn Rimes, Paul McCartney, Boxcar Willie, Ray Charles and others

Cajun Moon (J. J. Cale, 1974)

Dark Side of the Moon, album by Pink Floyd (1973)

East of the Sun (And West of the Moon) (Brooks Bowman, 1935)

Fly Me to the Moon (Bart Howard, 1954), by Johnny Mathis, Frank Sinatra, Tony Bennett and others

Harvest Moon (Neil Young, 1992)

Hijo de la Luna, by Spanish pop band Mecano (1986)

I Hear a New World – An Outer Space Music Fantasy (Joe Meek, 1960)

I Wished on the Moon (Dorothy Parker–Ralph Rainger, 1936), Billie Holiday

It's Only a Paper Moon (Harold Arlen–E. Y. Harburg–Billy Rose, 1933), Frank Sinatra

La Luna, album by Sarah Brightman (2000)

Love and the Moon, by Mel Tormé

Man on the Moon (R.E.M., 1992)

Moon Country (Johnny Mercer–Hoagy Carmichael, 1933)

Moon Love (Mack David–Mack Davis–André Kostelanetz, 1939), Glenn Miller

Moon on Bourbon Street, Sting

Moon River (Johnny Mercer–Henry Mancini, 1961), Andy Williams

Moon Song (Sam Coslow –Arthur Johnston, 1932)

Moonage Daydream (David Bowie, 1971)

Moonchild (Keith Jarrett)

Moondance (Van Morrison, 1970)

Moondance (Ernest Ranglin)

Moonlight and Roses (Ben Black–Neil Moret, 1925)

Moonlight Becomes You (Johnny Burke–Jimmy Van Heusen, 1942)

Moonlight in Vermont (Johnny Blackburn–Karl Suessdorf, 1943)
versions have been recorded by: Thomas Anders, Ray Anthony, Louis
Armstrong, Chet Baker, Tony Bennett, Acker Bilk, Stanley Black, Les
Brown and His Band of Renown, Jimmy Bruno, Charlie Byrd, Benny
Carter, Ray Charles, Rosemary Clooney, Nat King Cole, Sam Cooke,
Tommy Dorsey & his Orchestra, The Dorsey Brothers, Billy Eckstine,
Percy Faith, Ella Fitzgerald, Stan Getz, Dizzy Gillespie, Stephane
Grappelli, Buddy Greco, Johnny Hartman on his album *Songs from
the Heart*, Earl Hines, Billie Holiday, Stan Kenton, Frankie Laine,
Gordon Langford, Lataamon Laulajat (Finnish translation *Luumäen
kuun valo*), Kevin Mahogany, Johnny Mathis, Carmen McRae, Jane
Monheit, Nana Mouskouri, Gerry Mulligan, Willie Nelson on his
album *Stardust*, Oscar Peterson, Bucky Pizzarelli on his album *Doug
and Bucky*, Louis Prima, Linda Ronstadt, Neil Sedaka, Zoot Sims,
Frank Sinatra, Johnny Smith, Jo Stafford, Mel Tormé, Leslie Uggams,
Caterina Valente, Sarah Vaughan, Margaret Whiting, Bobby Womack

Moonlight Mile (Mick Jagger–Keith Richards, 1970), The Rolling Stones

Moonlight Mood (Harold Adamson–Peter de Rose)

Moonlight Serenade (Glenn Miller–Mitchell Parish, 1939), Glenn Miller

Moonlight Sinatra, album by Frank Sinatra (1966)
(Arrangements by Nelson Riddle)
All of the tracks feature the Moon:

Moonlight Becomes You

Moon Song

Moonlight Serenade

Reaching for the Moon

I Wished on the Moon

Oh, You Crazy Moon

The Moon Got in My Eyes

Moonlight Mood

Moon Love

The Moon Was Yellow (And the Night Was Young)

Moonshadow (Cat Stevens)

Mr. Moonlight (Roy Lee Johnson, 1962), The Beatles

No Moon at All (Redd Evans–Dave Mann, 1957)

Oh, You Crazy Moon (Johnny Burke–Jimmy Van Heusen), Mel Tormé, Frank Sinatra

Pink Moon, song and album by Nick Drake (1972)

Polka Dots and Moonbeams (Johnny Burke–Jimmy Van Heusen, 1940), Count Basie

Reaching for the Moon (Irving Berlin, 1930)

Reflection, by Tool (in which the moon reveals her secret: 'As full and bright as I am, this light is not my own'.)

Shine on, Harvest Moon (Jack Norworth–Nora Bayes, 1907)

Silver Wings in the Moonlight (Hugh Charles–Leo Towers–Sonny Miller, 1943)

Surfer Moon (Brian Wilson), on the Beach Boys' album Surfer Girl

That Old Devil Moon (Burton Lane–E. Y. Harburg, 1946)

The Moon Got in My Eyes (Johnny Burke–Arthur Johnston)

The Moon Was Yellow (And the Night Was Young) (Fred E. Ahlert–Edgar Leslie)

To the Moon and Back, by Australian duo Savage Garden

Walking on the Moon (The Police, 1979)

Books

These authors are generally considered to be the 'Big Three' in the science-fiction genre:

ARTHUR C. CLARKE
1917–2008

Clarke was probably most famous for his novel *2001: A Space Odyssey*, although many fans argue that this was not his greatest work. He also wrote the sequels *2010: Odyssey Two*, *2061: Odyssey Three*, and *3001: The Final Odyssey*. His novels *A Fall of Moondust*, *Earthlight*, *Rendezvous with Rama*, and *2001: A Space Odyssey* all feature colonies on the moon.

ROBERT A. HEINLEIN
1907–88

Heinlein was a science-fiction writer who wrote several books in which the moon features. The same characters and places appear in many of the books. Heinlein also worked on the films *Destination Moon* and *Project Moonbase*. His most famous series of books were the Future History series, the Lazarus Long series and the World as Myth series. There is a crater on Mars named after him. Heinlein was also the guest commentator with Walter Cronkite during the Apollo 11 moon landing.

ISAAC ASIMOV
1920–92

Asimov wrote extensively about the moon, including the fictional *The Tragedy of the Moon* and many works of theory.

BOOKS ABOUT THE MOON

2nd century AD: *Icaromenippus*, by Lucian of Samosta. A satirist who wrote fantastical tales. In this tale the hero flies to the moon, where he has adventures with the inhabitants of the sun and the moon. Incredibly, Lucian has been compared to Douglas Adams.

1502: *Orlando Furioso* (*Mad Orlando*), by Ludovico Ariosto. A collection of fantastical tales, including a trip to the moon.

1634: *Somnium* (*The Dream*), by Johannes Kepler. An Icelandic voyager is transported to the moon by aerial demons. Kepler wrote that mortals can reach the moon only with the help of moon demons. He describes gravity, temperature extremes, an exaggerated scale of topography and creatures of monstrous size.

1638: *The Man in the Moon or a Discourse of a Voyage Thither*, by Francis Goldwin. A Spaniard flies to the moon using a contraption pulled by geese.

1656: *The Other World or A Comical History of the Nations and Empires of the Moon*, by Cyrano de Bergerac.

1705: *The Consolidator*, by Daniel Defoe. The hero travels between China and the moon on an engine called *The Consolidator*. This was a satire on the English Parliament.

1865: *From the Earth to the Moon*, by Jules Verne. In this story a spaceship is launched from Florida, travels to the moon and returns to earth, landing in the Pacific Ocean. It is eerily similar to the route taken by the Apollo missions.

1887: *Les Exilés de la Terre* (*The Conquest of the Moon*), by Paschal Grousset. A Sudanese mountain made of iron ore is turned into a huge electro-magnet and catapulted to the moon.

1901: *First Men in the Moon*, by H. G. Wells. Cavorite, a gravity shield, is used to get a rocket to the moon.

1925: *Roverandum*, by J. R. R. Tolkien. The story was devised to distract Tolkien's son Michael who had lost his favourite toy, a lead dog. *Roverandum* is about a dog, Rover, who is turned into a toy. Rover goes to the moon and under the ocean to find wizard balls to get himself turned back into a normal-sized dog. The story was not published until 1998.

1928: *Doctor Dolittle on the Moon*, by Hugh Lofting. Doctor Dolittle travels to the moon and meets strange creatures, speaking plants and a prehistoric man who has grown into a giant.

1936: *Lost Paradise*, by C. L. Moore. A Northwest Smith story. The once-fertile moon is turned into an airless wasteland.

1951: *Prelude to Space*, by Arthur C. Clarke. Describes fictional events leading up to the first moon landing, which Clarke reckoned would happen in 1978.

1953 and **1954**: Two Tintin stories by Hergé are set on or about the moon: *Objective Moon* and *They Walked on the Moon*.

1955: *Earthlight*, by Arthur C. Clarke. The moon is caught in the crossfire of a war between earth, Mars and Venus.

1961–78: *Matthew Looney*, by Jerome Beatty Jr. A series of children's books featuring a government of moon dwellers trying to invade earth.

1977: *Inherit the Stars*, by James P. Hogan. First of the *Minervan Experiment* series. The moon is discovered to have previously orbited Minerva, which exploded 50,000 years ago to form the asteroid belt.

1989: *Hyperion*, by Dan Simmons. The moon is one of hundreds of colonies in space. Almost no one lives on it, as most of the other colonies offer a better environment.

1991: *Lunar Descent*, by Allen Steele. Set in 2024, the novel describes Descartes Station, a lunar base.

1993: *Assemblers of Infinity*, by Kevin Anderson and Doug Beason. Robots build a strange structure on the far side of the moon.

1993: *De Maan* (*Moon Handbook: A 21st-Century Travel Guide*), by Carl Koppeschaar. A German 'travel guide' to the moon.

1995: *Byrd Land Six*, by Alastair Reynolds. Features a lunar colony where Helium-3 is mined.

1996: *Transmigration of Souls*, by William Barton. Astronauts discover an alien moon base. The aliens have advanced technology, including teleportation and time travel.

1996 and **1997**: *Moonrise* and *Moonwar*, by Ben Bova. A private-equity-funded American lunar base rebels against the earth.

1998: *Moonfall*, by Jack McDevitt. The moon is threatened by collision with a comet as the first lunar base opens.

2002: *Ice*, by Shane Johnson. The marooned Apollo 19 crew discover an ancient but working lunar base.

2004: *Once in a Blue Moon*, a series of graphic novels written by Nunzio DeFilippis.

2008: *The Moon Book*, by Jimmy Joe Hughes. A handmade book of lunar miscellany, in full colour. Privately distributed and extremely rare. The only known existing copy is in private hands.

A Short Lunar Anthology

From 'Night'

The moon like a flower
In heaven's high bower,
With silent delight,
Sits and smiles on the night.

William Blake, 1757–1827

From *Villette* ('La Terrasse')

Where, indeed, does the moon look not well? What is the scene,
confined or expansive, which her orb does not fallow? Rosy or
fiery, she mounted now above a not distant bank; even while
we watched her flushed ascent, she cleared to gold, and in very
brief space, floated up stainless into a now calm sky.

Charlotte Brontë, 1816–55

From 'The Walrus and the Carpenter'
from *Through the Looking Glass*

The moon was shining sulkily,
Because she thought the sun
Had got no business to be there
After the day was done

Lewis Carroll, 1832–98

Fragment

It was a totally different moon than I had ever seen before.
The moon that I knew from old was a yellow flat disk, and
this was a huge three-dimensional sphere, almost a ghostly
blue tinged sort of pale white. It didn't seem like a friendly
place or a welcoming place. It made one wonder whether we
should be invading its domain or not.

Michael Collins, b. 1930
(quoted by Kevin W. Kelley, *The Home Planet*)

'At a Lunar Eclipse'

Thy shadow, Earth, from Pole to Central Sea,
Now steals along upon the Moon's meek shine
In even monochrome and curving line
Of imperturbable serenity.

How shall I link such sun-cast symmetry
With the torn troubled form I know as thine,
That profile, placid as a brow divine,
With continents of moil and misery?

And can immense Mortality but throw
So small a shade, and Heaven's high human scheme
Be hemmed within the coasts yon arc implies?

Is such the stellar gauge of earthly show,
Nation at war with nation, brains that teem,
Heroes, and women fairer than the skies?

Thomas Hardy, 1840–1928

From *The Iliad* (Book VIII, I. 687)

As when the moon, refulgent lamp of night,
O'er heaven's pure azure spreads her sacred light,
When not a breath disturbs the deep serene,
And not a cloud o'ercasts the solemn scene;
Around her throne the vivid planets roll,
And stars unnumber'd gild the glowing pole,
O'er the dark trees a yellower verdure shed,
And tip with silver every mountain's head

Homer, 8th century BC (translated by Alexander Pope)

From 'The Owl and the Pussycat'

They dined on mince, and slices of quince,
Which they ate with a runcible spoon;
And hand in hand, on the edge of the sand,
They danced by the light of the moon,
The moon,
The moon,
They danced by the light of the moon.

Edward Lear, 1812–88

Poem

Among the flowers, a jug of wine,
Alone, without companions, I drink from it,
I raise my cup to the shining Moon
We are three shadows, three persons.

Li Po, 701–62

From 'Endymion'

*There liveth none under the sunne, that know what to make of
the man in the moon.*

John Lilley, 1554–1606

From 'The Song of Hiawatha'

Saw the moon rise from the water,
Rippling, rounding from the water,
Saw the flecks and shadows on it,
Whispered, 'What is that, Nokomis?'
And the good Nokomis answered,

'Once a warrior, very angry,
Seized his grandmother, and threw her
Up into the sky at midnight;
Right against the moon he threw her;
'Tis her body that you see there.'

Henry Wadsworth Longfellow, 1807–82

From sonnet 'On his Blindness' (*Samson Agonistes*)

The sun to me is dark
And silent as the moon,
When she deserts the night
Hid in her vacant, interlunar cave.

John Milton, 1608–74

'The half moon shows a face of plaintive sweetness'

The half moon shows a face of plaintive sweetness
Ready and poised to wax or wane;
A fire of pale desire in incompleteness,
Tending to pleasure or to pain:
Lo, while we gaze she rolleth on in fleetness
To perfect loss or perfect gain.
Half bitterness we know, we know half sweetness;
This world is all on wax, on wane:
When shall completeness round time's incompleteness,
Fulfilling joy, fulfilling pain?
Lo, while we ask, life rolleth on in fleetness
To finished loss or finished gain.

Christina Rossetti, 1830–94

'To the Moon'

Art thou pale for weariness
Of climbing heaven and gazing on the earth,
Wandering companionless
Among the stars that have a different birth,
And ever changing, like a joyless eye
That finds no object worth its constancy?

Percy Bysshe Shelley, 1792–1822

From 'The Cloud'

That orbèd maiden, with white fire laden
Whom mortals call the moon

Percy Bysshe Shelley, 1792–1822

Astrophel and Stella, XXXI

With how sad steps, O Moon, thou climb'st the skies!
How silently, and with how wan a face!
What, may it be that even in heavenly place
That busy archer his sharp arrows tries?
Sure, if that long with love-acquainted eyes
Can judge of love, thou feel'st a lover's case;
I read it in thy looks; thy languisht grace
To me that feel the like, thy state descries.
Then, even of fellowship, O Moon, tell me,
Is constant love deemed there but want of wit?
Are beauties there as proud as here they be?
Do they above love to be loved, and yet
Those lovers scorn whom that love doth possess?
Do they call virtue there, ungratefulness?

Sir Philip Sidney, 1554–86

'Moon's Ending'

Moon, worn thin to the width of a quill,
In the dawn clouds flying,
How good to go, light into light, and still
Giving light, dying.

Sara Teasdale, 1884–1933

From 'Sonnet'

All night, through archways of the bridgèd pearl
And portals of pure silver, walks the moon.

Alfred, Lord Tennyson, 1809–92

From *From the Earth to the Moon*

There is no one among you, my brave colleagues, who has not seen the moon, or at least, heard of it.

Jules Verne, 1828–1905

Bibliography

The following is a list of some of the books that I used and found interesting while researching *The Book of the Moon*. There are of course many more.

Facts and Figures
The Lunar Source Book, edited by Grant Heiken, David Vaniman, Bevan M. French, Cambridge University Press, 1991.
A very technical book written by geologists, but it contains everything we know about the physical properties of the moon.

To a Rocky Moon, by Don E. Wilhelms, University of Arizona Press, 1993.
An interesting and informative book, written by a geologist.

The Once and Future Moon, by Paul D. Spudis, Smithsonian, 1996.
Another very interesting description of the moon, written by a geologist.

Astronomers
Mapping and Naming the Moon, by Ewen A. Whitaker, Cambridge University Press, 1999.
The definitive book about mapping the moon.

Astronomy: A Popular History, by J. Dorschner *et al*, Van Nostrand Reinhold & Co., 1975 edition, Leipzig.

Observing the Moon: A Modern Astronomer's Guide, by Gerald North, Cambridge University Press, 2000.
A comprehensive book with a useful bibliographical element.

Observing the Moon, by Peter T. Wlasuk, Springer, 2000.
An excellent guide to observing the moon. The book comes with a CD.

Gods and Myths
A Lunar Almanac, by Rosemary Ellen Guiley, Cynthia Parzych Publishing Inc., 1991.
A lunar anthology with interesting chapters on myths and medicine.

The Sun Maiden and the Crescent Moon, by James Riordan, Canongate, 1989.
A collection of Siberian folk tales.

The Book of the Moon, by Tom Folley, Greenwich Editions.
A fascinating meander through moon culture and myths.

Gardening and the Weather
Gardening & Planting by the Moon, by Nick Kollerstrom, W. Foulsham and Co.
A very good guide to lunar gardening, with calendar. Published annually.

Agriculture, by Rudolf Steiner, Biodynamic Farming and Gardening Association Inc., 1993.
This is the seminal work on biodynamic agriculture.

Gardening for Life, by Maria Thun, Hawthorn Press, 1999.

Stella Natura Calendar, Camphill Village, Kimberton Hills, PA 19442.

Astro Calendar (Northern Hemisphere), Growing Biodynamic LLC, Kimberton Hills, PA 19442.

Astro Calendar (Southern Hemisphere), by Brian Keats.

Astronauts, Cosmonauts and Lunar Exploration
Exploring the Moon, by David M. Harland, Springer published with Praxis Publishing, 1999.
A very comprehensive account of the Apollo programme.

Apollo: The Definitive Source Book, by Richard W. Orloff and David M. Harland, Springer published with Praxis Publishing, 2006.
This contains all the information on Apollo that you could possibly want.

A Fire on the Moon, by Norman Mailer, Pan Books, 1970.
In my opinion, the best account of the Apollo 11 mission ever written.

The Dark Side of the Moon, by Gerard DeGroot, Jonathan Cape, 2007.

Moondust, by Andrew Smith, Bloomsbury, 2005.
The biographies of the moonwalkers.

Magic, the Occult, Astrology, Alchemy, Prophecy, Fortune-Telling, Spells and Superstition
Moon Wisdom, by Sally Morningstar, Southwater.

Magick & Rituals of the Moon, by Edain McCoy, Llewellyn Publications, 2001.

A Brief History of Ancient Astrology, by Roger Beck, Blackwell Publishing, 2007.

The Moon in Occult Lore, by Corinne Heline, New Age Press, 1970.

Moon Magic, by Lori Reid, Carlton, 1998.

The Wicca Handbook, by Eileen Holland, Robert Hale, 2000.

Power and Knowledge, by Tamsyn Barton, University of Michigan, 1997.
An interesting and sceptical book about magic and astrology.

Medicine, Madness, Werewolves and Science
Moon Madness, by Paul Katzeff, Robert Hale, 1990.

The Lunar Effect, by Arnold L. Lieber, MD, Corgi, 1979.
A scientist attempts to analyse the moon's influence and to prove its effect on our world.

Astrology in Medicine, by C. A. Mercier, Macmillan, 1914.
A series of satirical lectures to the Royal College of Physicians.

The Werewolf, by Basil Copper, Robert Hale, 1977.

Miscellany
Earth–Moon Relationships, by Cesare Barbieri and Francesca Rampazzi, Kluwer Academic Press, Dordrecht, 2001.

The Skeptic's Dictionary, by Robert Todd Carrol, John Wiley and Sons, 2003.
Includes good debunking articles about the moon and lunar exploration.

The Moon: A Biography, by D. Whitehouse, Headline Review, 2002.

Picture Acknowledgements

Endpapers [hardback edition only]
Front: *Tabula Selenographica … Helvelius quam Riccioli* by J. Doppelmayr and J. B. Homann, *c.*1740: Jonathan Potter Limited/www.jmaps.co.uk
Back: *The Palace of the Queen of the Night*, set design for W. A. Mozart's *The Magic Flute* by Karl Friedrich Schinkel for a production in Berlin, 1816, Deutsches Theatermuseum, Munich: © INTERFOTO Pressebildagentur/Alamy

In the text
Figures on pages 17, 19, 20, 23, 26/27, 35, 40, 41, 59, 61, 82, 83, 161, 162 and 163 are by HL Studios
Frontispiece, 8/9, 10/11: fotolia
37 *A Group of Lunar Craters SE of Tycho*, engraving after James Nasmyth, 1863: Wellcome Library, London
43 (& 12) *The moon, viewed in oblique sunlight*, stipple engraving by James Russell, 1806: Wellcome Library, London
54 Tycho Brahe, title page of *Astronomiae instauratae mechanica*, 1602: Science Museum Library
62 Stonehenge, 18th-C engraving: © The Print Collector/Alamy
64 Tycho's quadrant from *Astronomiae instauratae mechanica*, 1602: Wellcome Library, London
65 Armillary sphere, frontispiece from *A Plain Treatise of the First Principles of Cosmographie* by Thomas Blundeville, 1594: private collection/ The Bridgeman Art Library
71 Ptolemy and Astronomy from *Margarita philosophica* by Gregorius Reich, 1503: Wellcome Library, London
77 Top: Nicolaus Copernicus, woodcut by Tobias Stimmer, mid-16th C: © INTERFOTO Pressbildagentur/Alamy
77 Btm: Galileo Galilei, frontispiece from his *Il Saggiatore*, 1623: Science Museum Library
78 Johannes Kepler, engraving, 17th C: Science Museum
89 Photo of the moon by Dr Henry Draper, 1885: Detlev van Ravenswaay/Science Photo Library
93 (& 54) Moon map by Claude Mellan, 1635: Science Photo Library
95 Moon map by Johannes Helvelius, from *Selenographia, sive Lunae descriptio*, 1654: Royal Astronomical Society/Science Photo Library
96 Moon map by Giovanni Battista Riccioli, from *Almagestum Novum*, 1651: Royal Astronomical Society/Science Photo Library
99 Map of the SE quadrant of the moon from *Selenographia* by Wilhelm Madler based on observations by Johann Heinrich Beer, 1834: Science Museum Pictorial
101 *The Maiden in the Moon*: Royal Astronomical Society/Science Photo Library
114 Sun and Moon from *Nuremberg Chronicle* by Hartmann Schedel, 1493: Dr Jeremy Burgess/Science Photo Library
125 *Chang'e Flies to the Moon*, book illustration, 1920s: © Lordprice Collection/Alamy
127 Hecate: © The Print Collector/Alamy
130 Moon mask, collected 1893, Inuit, Andreofsky, West Alaska: Werner Forman Archive/Sheldon Jackson Museum, Sitka, Alaska
136 Luna by Alfred Kubin, from *Die Planeten*, 1943: private collection/The Bridgeman Art Library

137 *The rabbit in the moon* from a history of the Aztecs and the conquest of Mexico, Spanish: Ms 219 f.223v, Biblioteca Medicea-Laurenziana, Florence,/ The Bridgeman Art Library
140 *Amaterasu, the sun goddess, holds moon that contains a hare*, 1407, Japanese: Ronald Sheridan © Ancient Art & Architecture Collection Ltd
156 (& 152) Pliny the Elder from *Nuremberg Chronicle* by Hartmann Schedel, 1493: © World History Archive/Alamy
178 Neil Armstrong on the surface of the moon, 20 July 1969: NASA
181 Goddard's rocket, 16 March, 1926: NASA
182 Top left: Werner von Braun with a model of a V2 rocket: NASA
Top right: Werner von Braun in front of Saturn IB launch vehicle, January 1968: NASA
Btm: Saturn V diagram: NASA
190 Far side of the moon, 7 October 1959: RIA Novosti/Science Photo Library
191 (& 178) First close-up photo of the moon's surface, 3 February 1966: Jodrell Bankk/Science Photo Library
194 Top: Luna-9 automatic lunar station, 1966: RIA Novosti
Btm: Artwork of the Luna-9 space probe, 1966: RIA Novosti
199 Artwork of the Luna-17 space rover, 1970: RIA Novosti
203 Artwork of the Luna-24 space probe, 1973: RIA Novosti
207 Ranger spacecraft, 1961: NASA
212–257 All Apollo mission photos: NASA except for 246 Apollo 17 crew: NASA/Science Photo Library
258 Background: Phases of the moon, engraving from *The Selenic Shadowdial or the Process of the Lunation* by Pierre Miotte, mid-16th C: Wellcome Library, London
260 *Wizard Consulting the Moon and the Stars*, woodcut from a collection of chapbooks on esoterica, English: private collection/The Stapleton Collection/The Bridgeman Art Library
261 Astrologers at a birth, 16th C: INTERFOTO Pressbildagentur/Alamy
264 Phases of the moon from Galen's *Opera Varia*, mid-14th C: WMS 286 f. 78, Wellcome Library, London
268 Lilith, after a painting by Kenyon Cox, 1892: © Mary Evans Picture Library/Alamy
270 Alchemical miniature, 16th C, Dresden: Wellcome Library, London
274 Queen Luna: Fortean Picture Library
275 (& 258) *Dr Faustus in his Study*, Rembrandt, 1652, etching: © The Print Collector/Alamy
278 Witches, *c.*1600, Mary Evans Picture Library
280 Grimoire: Fortean Picture Library
281 Background: Pentangles: Fortean Picture Library
283 Saxon idol of the moon, 1834, wood engraving: © eaglecrown
287 Illustration from *Projection of the Astral Body* by Hereward Carrington and Sylvan Muldoon, 1929: © Mary Evans Picture Library/Alamy
290 Tarot card of the moon: © Mary Evans Picture Library/Alamy
303 Claudius Galen, engraved portrait: © Mary Evans Picture Library/Alamy
305 Title page of Hippocrates' *Opera Omnia*, 1526: Wellcome Library, London
305 (& 300) *Four humours*, engraving from *Quinta essential* by L. Thurneysser, 1574

309 Democritus, detail of an engraving by Wenceslaus Hollar after Rembrandt, 165?: Wellcome Library, London

313 (& 300) 'Celestial Influences on Men, Animals and Plants' from *Liber divinorum operum simplicis homonis* by Hildegard von Bingen, *c*.1200, this edition 1928: Wellcome Library, London

314 Title page of *Recetario de Gallieno* by Claudius Galen, 1518: Wellcome Library, London

315 Bloodletting, manuscript illumination: Wellcome Library, London

317 *Moonstruck women*, 17th-C Dutch engraving: Science Photo Library

323 *Werewolf attacking children*, German engraving, *c*.1500: Mary Evans Picture Library

324/5 Background: Man attacked by a werewolf, 1517: © Mary Evans Picture Library/Alamy

338 Top: 'A View of the Inhabitants of the Moon', *New York Sun*, August 1835, from a series of articles wrongly attributed to Sir John Herschel: Mary Evans Picture Library

338 Btm: Sir John Frederick William Herschel by Julia Margaret Cameron, 1867: Wellcome Library, London

340 Footprint on the moon, 20 July 1969: NASA

342 Apollo moon landing: NASA

344 'Man's Reach', a 1968 Herblock cartoon: copyright by The Herb Block Foundation

347 *Trip to the Moon*, 1902: Méliès/The Kobal Collection

356 'The Owl and the Pussycat' illustration by L. Leslie Brooke to Edward Lear's *Nonsense Songs*, 1871: © Lebrecht Authors

360 Illustration by Émile Bayard to Jules Verne's *Autour de la Lune*, 1870: © Mary Evans Picture Library: Alamy

Colour

First section

Face of the Moon, pastel by John Russell, 1793–97: © Birmingham Museums and Art Gallery/The Bridgeman Art Library

Blanchard bone, 32,000–25,000 BC, musée d'archéologie nationale, Saint-German-en-Laye, Dordogne: © RMN/Loic Hamon; calendar circle, Nabta Playa, Egypt, 5th millennium BC: © Mike P. Shepherd/Alamy; cave painting, 14th millennium BC, Lascaux, Dordogne: Sisse Brimberg/National Geographic Stock; standing stones, Callanish, Western Isles, 2000 BC: © Adam Woolfitt/Robert Harding World Imagery/Corbis

Thoth, relief, Luxor, Egypt, 11th C BC: Jon Bodsworth; Egyptarchive; Selene, fresco, House of Ara Maxima, Pompeii, 1st C AD: Bildarchiv Steffens/Ralph Rainer Steffens/The Bridgeman Art Library; chariot of the moon, detail of a page from Cicero's *Aratus*, 1125–50: Cotton Tiberius B. V, part 1, f. 47/© 2008 The British Library; colossal stone head of Coyolxauhqui, 1300–1521 AD, Aztec: Werner Forman Archive/National Museum of Anthropology, Mexico City; moon mask, 19th/20th C, Kwakwaka'wakw culture, NW coast Canada: Werner Forman Archive/Provincial Museum, Victoria, British Columbia, Canada

Astronomer, from an 18th-C copy of a 13th-C work by the Armenian philosopher Hovhannes Erznkac'i: Or (Armenia) 7/ Wellcome Library, London; astronomers , 17th-C Ottoman miniature: The Art Archive/University Library istanbu/Gianni Dagli Orti; installation of an armillary globe, 16th-C Ottoman miniature from an astronomical treatise: The Art Archive/University Library istanbu/Gianni Dagli Orti; diagram from the *Miscellany* of Iskandar Sultan, 1410–11, southern Iran: Add. 27261, f.410/© 2008 The British Library; Jantar Mantar, Jaipur, Rajasthan: © Robert Preston Photography/Alamy

Diagram of Meteorology, drawn and engraved by John Emslie, 1846: Science Museum Pictorial; *Herschel Table of the Weather*, 1815: Science Museum Pictorial

Second section

Apollo mission badges: all courtesy NASA Images

'The anatomy of Man and Woman' from the *Très Riches Heures du Duc de Berry* by Pol de Limbourg, 15th C: Musée Condé, Chantilly, France/Giraudon/The Bridgeman Art Library; Diana and her followers, from the works of Christine de Pisan, *c*.1410–15: Harl 4431 f.101/British Library, London, UK/© British Library Board. All Rights Reserved/The Bridgeman Art Library; Luna from a calendar for the year 1446, German: Add. 17987, f. 75v/© The British Library; Luna from an edition of *De Sphaera*, 15th C, Biblioteca Estense, Modena: Fortean Picture Library; magi measuring the moon, tarot card: Photos12.com –ARJ; 'The Moon and You', *Fate Magazine* cover, 1954: Mary Evans Picture Library

Alchemical emblems from *Alchemical Receipts*, 15th C, England: Harley 2407, f.68/© The British Library; young king in a glass vessel, detail of a page from *Splendor Solis* by Salomon Trismosin, 1582, Germany: Harley 3469, f.29/© The British Library; Hermes Trismegitos, engraving from *Chymisches Lustgärtlein* by Daniel Stolzius, 1624: © INTERFOTO Pressbildagentur/Alamy; emblematical drawing from *On the Philosopher's Stone*, 17th C: Sloane 1316, f. 10/© The British Library; alchemical illustration from *Speculum Philosophiae* by John Dustin, 17th century, England: Sloane 2480 f. 2/© 2008 The British Library

Scene from *The Adventures of Baron Munchausen* by Rudolph Erich Raspe, *c*.1850: Ann Ronan Picture Library/Heritage Image Partnership; 'Blue Moon' by Richard Rogers and Lorenz Hart, sheet music, 20thC: © Vintage Image/Alamy; poster for *Cat-Women of the Moon*, 1953: Astor Prod./The Kobal Collection; *2001: A Space Odyssey*, 1968: MGM/The Kobal Collection; *A Grand Day Out*: © NFTS 1989

City on the moon, artwork: Chris Butler/Science Photo Library; robot mining machine, artwork: University of Wisconsin Center for Space Automation and Robotics/Science Photo Library

Index

Royal Society, London 156, 316

Russell, John: lunar maps 99

Russia/Soviet Union 181–2, 183, 186, 187
myths 145–6
see also Luna programme

Sardnuna (goddess) 119

Saunder, Samuel 103, 104

Saxby, Stephen Martin 175

Schirra, Walter M. Jr 213, 217, 218

Schmitt, Harrison ('Jack') 213, 246, 247–9, 256

Schrader, Johann Gottlieb: telescope 98

Schrödinger Basin 53

Schröter, Johann Hieronymus 98 Selenotopographische Fragmente 98

Schweickart, Russell R. 213, 221, 222

Scott, David R. 37, 213, 221, 222, 240, 241–2, 249, 254

sea slugs 330

sea urchins 329

'seas', lunar see Maria

seeds, sowing 166, 167

Selene (goddess) 119, 128, 139, 294

Serenitatis Basin 30, 48, 53

sextants 65

Shelley, Percy Bysshe: 'To the Moon' 358
'The Cloud' 359

Shepard, Alan B., Jr 213, 237, 238–9, 251, 252–3

Shorty (crater) 51, 248–9

Siberian myths 145–6

Sibrel, Bart 251

sidereal cycle 161

sidereal months 21

Sidney, Sir Philip Astrophel and Stella 359

silica 164

Sin (god) 114, 115, 119, 131–2

Sina (goddess) 134

Sinag (god) 119

Sirdu (A) (goddess) 119

solar nebula 25

Soma (god) 119, 129

sooty terns 329

South Pole Basin 53

sowing seeds 166, 167

Soyuz 1 187

spectroscopy 75–6

spells and spell casting 276, 279, 280, 281, 289
for better sex 286
for leaving someone 286
plants for 287
preparing environment for 285
for prosperity 285–6
rules 284–5

spherical aberration 83, 85

Sputnik 186

Stafford, Thomas P. 213, 222, 223

Steiner, Rudolph 157, 158

Steno (crater) 248

Stonehenge, 62

Stubb, Peter 325

Sudines 67

suicide 332

superstitions 295–9, 304

Surveyor programme (US) 187, 207, 208–10

Swigert, John L., Jr 213, 234, 235–6

synodic cycle 161, 162

synodic months 21

System of Lunar Craters, The 105, 107

syzgy 19

Talbot, William Fox: The Pencil of Nature 88

Tanit (goddess) 119, 133

tarot 289–90

Taurus-Littrow 53

Teasdale, Sara: 'Moon's Ending' 359

tectonic activity 48

Tecuciztecatl (god) 119, 122–3

telescopes 75, 77, 79, 81–8, 176

Tennyson, Alfred, Lord: 'Though night hath climbed her peak' 359

Thales 69

Theophilus (crater) 51

Theophrastus: A History of Physics 173, 175, 176

Thoth (god) 114–15, 119, 125, 126, 271

Thun, Maria 158

Tiazolteotl (goddess) 123

tides 18–19, 332, 333

and movement of animals 329

Times Atlas of the Moon, The 111

Titania (goddess) 119, 136

toads 291

Torricelli, Evangelista 174

Tranquillity Base 53

transient lunar phenomena (TLP) 46

transplanting 167, 170

trees
felling 154–5
grafting 167
pruning 167, 170
transplanting 167, 168

Trivia (goddess) 127, 136

Tsukuyomi (god) 131

Turner, Herbert H. 104

al-Tusi, Nasir al-din 73

Tycho (crater) 31, 33, 34, 41, 47, 51, 248

Ulugh Beg 65, 67

United States of America 180–81, 183, 184, 186, 187, 206
see also Apollo missions; Lunar Orbiter; Native Americans; Pioneer; Ranger

Ursula 119

V2 rockets 182

vegetables 171–2
see also biodynamic farming

Verne, Jules 360

video cameras 88–9

volcanic activity 42–6

volcanic gas 43

volcanic lava 30, 33, 36, 43, 44–5

Voyager 1 187

Wan Hu 180

waning moon 20

water (on moon's surface) 24, 45

water, moon 282

waxing moon 20

weather
forecasting 173–6
superstitions 298–9
traditional lore 177

weed control, biodynamic 168

werewolves 322, 324, 325–7

West, Lynette 166

Whitaker, E. A. 108

Whitaker, E. W.: Photographic Lunar Atlas 105, 108

White, Ed 213, 214, 215

Wicca 279

witchcraft 278

wolves 291
see also lycanthropy; werewolves

Worden, Alfred M. 213, 240, 241, 242

worms: and responses to moon 328–9

Wren, Sir Christopher: lunar globe 100

Yellow Woman (goddess) 119

Yerkes Observatory 84, 87, 105

Yolkai Estasan (goddess) 119, 120

Young, John W. 213, 222, 223, 243, 244–5, 255

Yue-Lao 138

Zarpandit (goddess) 119, 132

Zirna (goddess) 119, 126

zodiac, the 262, 263, 264
and plants 159–61

Zoroastrianism 274